JN028063

英語と日本語で学ぶ
知覚情報科学

Cognitive Informatics in English and Japanese

Ruggero Micheletto
戸坂 亜希
及川 虎太郎
著

共立出版

Foreword

This book has been written for the students of Yokohama City University as support for the course of "Cognitive Information Science". It is an original course that stays in the middle ground between physics/mathematics and biology/brain science. The book wants to answer these questions: how does a living organism work? How the various body parts and organs communicate to each other? What is the brain? Is it the CPU of an organism? These are fundamental questions that do not have a clear answer yet. However, there is a huge number of textbooks that approach the detailed description of human organs, cells and the nervous system.

In this book we intentionally want to keep on the surface, we avoid to go down into the biology, the chemistry or the physiology of an organism. We want to look from above and try to grasp the big picture: the human body is a machine. It works very similarly to a robot, it exchanges electric signals and it is controlled by a computational central unit: the brain. In the first chapters of this book, we describe the so called "five senses", sight, hearing, smell, taste and touch. The approach is from the physicist/engineering point of view: what kind of signals are generated and by what mechanism? We describe the basics of a neuronal cell and its unique signal, the "spike". In the central part of the book we explain the main principles that govern the brain reactions, and we describe the most modern concepts of intelligence. What is intelligence and how it is defined? We want to make intelligent machines, right? So, in the last chapters of the book we give the basis of modern artificial intelligence. We introduce fundamental models of the neuron and the most basics artificial intelligence element: the perceptron.

Not only we describe the system, but we also give programming examples that implement the elements of these AI topics, so the students can test those ideas by themselves. Finally, we give other examples of more complex neural networks up to the most recent hierarchical temporal memory (HTM) networks.

I want to thank the co-authors Aki Tosaka and Kotaro Oikawa for their help with the Japanese part of the book. They didn't only help in the translation, they actually participated to the development of the whole book with sharp suggestions, rewriting and making fundamental improvements. A big thank goes specially to A. Tosaka for her help in the development of the first chapters and the revision of the whole book.

Chapter 7 has been partially inspired by the author's participation to research activities with prof. Jeremy Wolfe. Very helpful were his keen suggestions and the support of researchers in his laboratory[1)]. A special thank goes to Kris Ehinger[2)] that gave a fundamental contribution to key parts of the chapter.

Ruggero Micheletto
April 2020

1) Visual Attention Lab, Harvard Medical School, 65 Landsdowne Street, Suite 404A, Cambridge, MA 02139

2) Formerly at Visual Attention Laboratory, now Senior Lecturer in the School of Computing and Information Systems at the University of Melbourne, Grattan Street, Parkville, VIC 3010, Australia

はじめに

街中でふいに懐かしい香りがして，今はもう会えない人を思いがけず思い出したことはありませんか？　それは嗅覚という人間の知覚が，脳の機能である記憶と複雑に結びついているためです．知覚とは，生きるために生物に備わっている大切な機能です．栄養をとるため，危険を回避するため，種を増やすため，知覚はそのセンサーとして使われます．そしてそのセンサーから得た情報は，脳によって処理されます．

近年，人工知能，AIという言葉はあちこちで頻繁に使われています．そして多くの研究者によって研究が盛んに行われています．冒頭に書いたように，懐かしい香りによってふいに昔の記憶が呼び起こされるということを人工知能によって再現するのは難しいように思われます．

しかし，技術はどんどん進歩しています．人間の脳の機能を凌駕する究極の人工知能が生まれる日も，遠くはないのかもしれません．けれども，究極の人工知能とは，どういうものでしょうか．人工知能の機能を追求し，活用すると，私たちの暮らしは豊かになるのでしょうか．実のところ，私にはまだよくわかりません．しかし，その究極の人工知能の実現のためには，まず人間の知覚と脳の機能を知らなければならないということはわかります．

本書では，人間の五感にはじまり，その五感から情報がどのように伝達され，どのように処理されていくのか？　という疑問に対して，知覚情報科学というキーワードを軸にまとめられています．本書を通じて，読者の皆さんに知覚情報科学という分野を知ってもらい，今後の展望に興味を持っていただければ幸いです．

今回，本書を翻訳することで，これまで知らなかった知覚についての基礎知識，脳の機能，そしてその脳の機能をどのように再現できるかといったことについて学びました．本書は，2015年に出版した『英語と日本語で学ぶ熱力学』

と同様に，知覚情報科学について，英語と日本語の両方で学べる本です．左側に英語，同じ内容の日本語が右側に書かれています．

また，本書後半では良く知られた学習系アルゴリズムのコードの概要やアプリケーションを複数紹介しています．本書の特徴である英語と日本語の併記による多視点的な理解に加え，実際にパソコンに向かって作業をすることで，より一層理解が深まることを期待します．

2020 年 4 月吉日

戸坂亜希・及川虎太郎

Contents

目　次

This book gives a "big picture" view of the functioning of the brain with a novel educational approach.

Chapter 1

Information pathways: the five senses

1.1 Introduction

A biological entity like a human being or a complex animal can be considered as an advanced information system. In a nutshell, an animal receives information, processes it and reacts to that with other information.

A simple bacterium receives information in the form of chemical signals, temperature conditions or even electrical stimuli. It somehow combines these, and reacts accordingly.

A complex animal, compared to a bacterium, performs deeper elaboration of multiple signals. However, from the informatics point of view, it is still a system that combines some kind of information and reacts to it according to certain rules. The reaction can trigger other kinds of information, such as sounds, release of chemicals, change in temperature. Even movement can be interpreted as information.

In this framework, let's study how the human body works. We will give hereafter an outline of the main information pathways that exist in the body of a human being. First of all, we have to understand what is a cell and how it processes signals.

1.2 The cell

A cell is the main constituent of a biological system. The human body is made by about 10 to 50 trillion (10–50 10^{12}) cells. There are hundreds of different types of cells in human bodies, and the majority are fat and muscle cells. Here we will focus on the functioning of the so-called *excitable*

第1章

情報の伝わり方：五感

1.1 はじめに

人間や複雑な構造をもつ動物のような生物学的な存在は，高度に進化した情報システムと見なすことができる．簡単に言えば，木の実の殻から情報を受け取り，データを処理し，他の情報にたどり着くことができるのが動物である．バクテリアのような単純な生物であっても，化学的な信号や温度の状態，さらには電気的な刺激といった情報を受け取ることができる．そして，どうにかして，その情報に応じた反応をする．

バクテリアに比べて複雑な動物は，複数の信号を組み合わせてより精密な動きをすることができる．しかしながら，情報科学という観点からみれば，それは単に何かしらの情報を集めて一定のルールに則って反応するシステムでしかない．反応は他の種類の情報を引き起こすことが可能である．例えば音，化学物質の放出，温度の変化，または運動といったことが情報の一種として判断される．

この章では，人間の体がどのようにはたらくのかを学ぼう．まずは人間の体の中に存在するおもな情報経路の概要を説明する．はじめに，細胞について，また細胞がどう信号を伝えるのかについて学ぶ．

1.2 細胞

細胞[1]は，生物学上の組織を構成する主たる構成物である．人間の体は，10～50兆[2]個（$10\sim50\times10^{12}$個）もの細胞でできている．細胞には数百もの違

1) 細胞：cell.
2) trillion は通常1兆.

cells. Those are cells that receive and transmit information in a body. The electro-chemical pathways by which a cell functions are very complex and are studied by biologist and physiologists in details. However, here we want to study only the higher level functionalities of a cell, so we will neglect the detailed chemical processes occurring in the cells. We only need to know that cells receive stimuli and respond to them with an electric signal. This signal is reproducible even in different conditions, has a characteristic shape and it is generally called *action potential* or *spike*.

You may ask: how can a simple cell act like and electronic device and produce electrical signals? To understand this, we have to describe few fundamental things about a cell.

A cell is a wet roundish object, immersed in a watery media composed by various substances. These substances include several types of ions, so the media is conductive. The cell's external membrane, that surrounds the cell's interior organs, is made of *phospholipids*, a bilayer of soft, insulating matter. This membrane is not perfectly hermetic, there are places where water and ions can circulate. These places are called *gates*, *ion channels* or *ion pumps* (naming depends on the context of discussion, figure 1.1).

There are many types of channels depending on the kind of ions that are allowed to go through. Most important are the sodium (Na) channels and potassium (K) channels. These molecular devices *open* or *close* at random intervals, with a determined probability that depends on certain conditions.

For example, the probability of a channel to open may depend on the difference of voltage between the cell's inner body and the outside. Or, the probability to open may depend on the external potential, cell currents or other electro-mechanical conditions.

The pumps, instead utilize energy from the cells to pump in or out ions. The so called *sodium/potassium pump* consumes adenosine triphosphate (ATP) and constantly maintains an electric imbalance between the cell interior and exterior[Karp, 2008]. This imbalance is the main reason why,

う種類があり，人間の体のおもな細胞は脂肪細胞と筋細胞である．ここで，いわゆる**興奮性細胞**の機能に着目したい．興奮性細胞とは体のなかで情報を受け取り，そしてそれを伝達するはたらきをもつ細胞である．細胞にはたらく電気化学的な経路はとても複雑であり，生物学者や物理学者はその詳細についての研究を行なっている．しかしながら，ここでは高いレベルにおける細胞の機能性について学びたいので，細胞のなかでの化学的な過程についてはあまり考えないことにしよう．ただ，細胞がどうやって刺激を受け，電気信号を発生することで反応するといったことを知ればよい．この信号は異なる状況下においても再現が可能であり，特徴的な形をもち，一般的には，**活動電位**や**スパイク**[3]と呼ばれる．

あなたはおそらく「どのように単純な細胞が，まるで電子デバイスのようにはたらき，そして電気信号を生み出すことができるのだろうか？」といった疑問を抱くであろう．これを理解するためには，まず細胞について，以下に記すような基礎的なことを学ぶ必要がある．

細胞は水分を含み丸い形をしていて，そして様々な物質からなる水のような媒体のなかに存在する．この物質は何種類かのイオンを含み，そして導電性をもつ．細胞外側の細胞膜は細胞内部の器官を包んでいて，**リン脂質**[4]からなる二重層でできており，柔らかく，絶縁体である．細胞膜には完全な密閉性はなく，水やイオンなどが通過することができる場所がある．この場所のことを，**イオンチャネル**[5]や**イオンポンプ**[6]といい，その役割によって呼び分ける (図 1.1)．

イオンの種類によってそれぞれを通り抜けることができる様々な種類のイオンチャネルがある．最も重要なものは，ナトリウム (Na)[7]チャネルとカリウム (K)[8]チャネルである．これらの分子デバイスは，ランダムな間隔でその経路を開閉するが，その確率というのは，それぞれの状況に依存する．

---

3) スパイク：電気的な信号が急激な上昇の後に急激に降下することをスパイクと呼ぶ.
4) リン脂質：phospholipids.
5) イオンチャネル：ion channels.
6) イオンポンプ：ion pump.
7) ナトリウム：sodium.
8) カリウム：potassium.

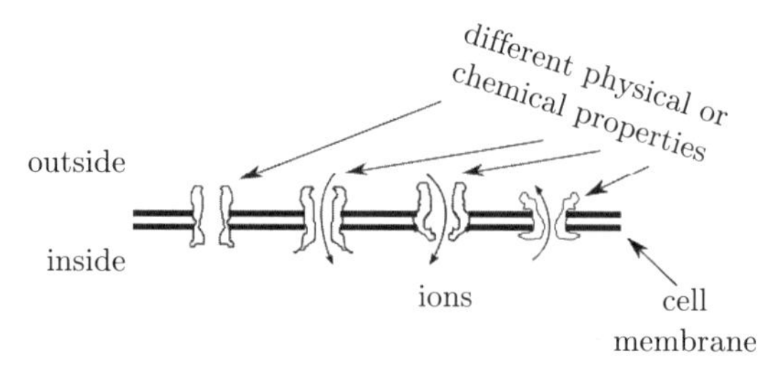

Fig. 1.1 The cell's membrane and some gates. Gates respond to physical, mechanical or chemical stimuli opening or closing the ion flux.

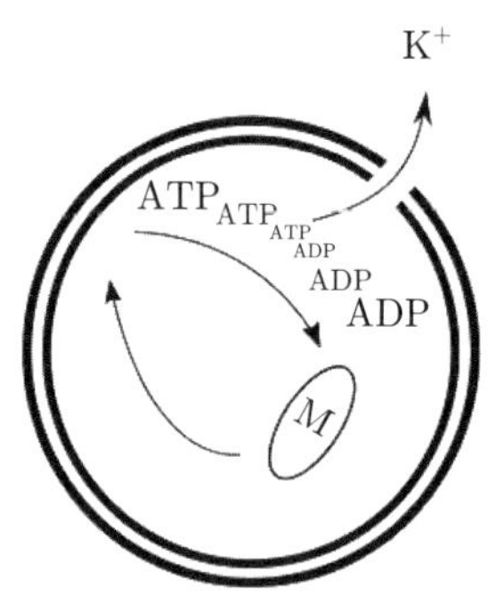

Fig. 1.2 A sketch of a cell, with an ion pump channel. The ion pump maintains a difference of ion concentration between the internal and external part of the cell. This results in a difference of potential. The ion pumps require energy that is obtained transforming ATP to ADP (represented in the picture by the characters). New ATP molecules are generated in the mitochondria organ, which is shown as the ellipse marked with M in the figure above.

when a channel opens, ions circulate and generate currents in certain conditions (figure 1.2). For example, an ion channel can act as a *receptor site*. This means that when a particular chemical reaction occurs at the channel location, the channel probability to open is increased and this results in a current that may lead to a sequence of electrical and chemical events that produce a characteristic electric signal (a *spike*) or *action potential*.

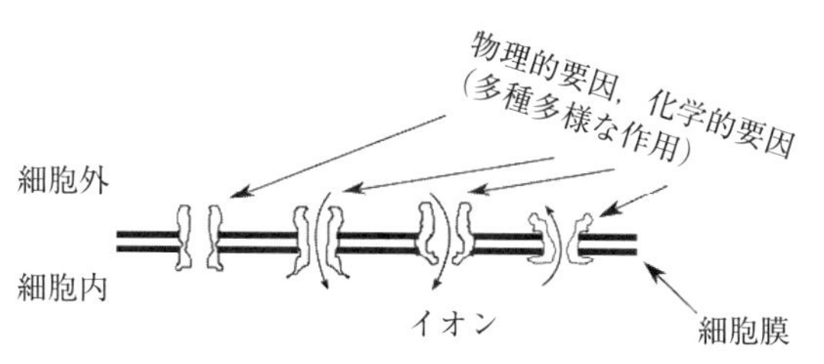

図 1.1 細胞膜といくつかのゲートの模式図．ゲートは物理的，機械的，化学的な刺激で開閉してイオンを出し入れする．

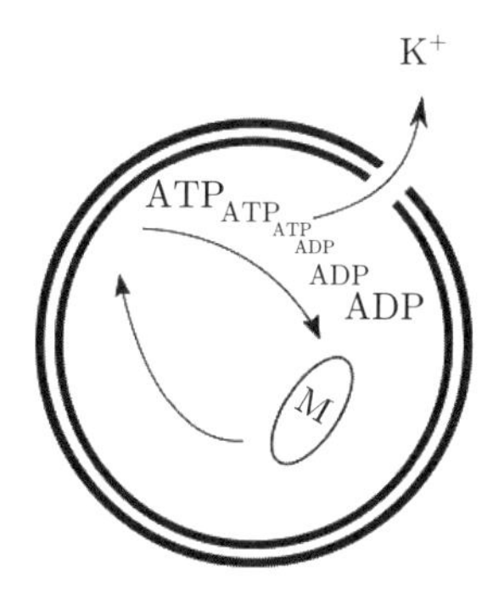

図 1.2 イオンポンプをもつ細胞の模式図．イオンポンプは細胞の内部と外部のイオン濃度の差を保つ．その結果，ポテンシャルの差を生み出す．イオンポンプは ATP を ADP に変換するときに得られるエネルギーを必要とする．(図中に文字で表現) 新しい ATP 分子は，図中に楕円 (M) で示したミトコンドリアによって生成される．

例えば，チャネルが開く確率は，細胞の中と外の電圧の差によって決まる．あるいは，外部のポテンシャル，細胞の中の複雑な現象[9]や他の電気的機械的な状況に依存する．

チャネルではなく，ポンプは細胞にイオンを入れたり出したりして生成したエネルギーを利用する．このナトリウム・カリウムポンプはアデノシン三リン酸 (adenosine triphosphate：ATP) を消費し，細胞の内側と外側での電荷の不均衡を保つ．この不均衡は，チャネルが開くと，イオンは循環し，ある一定の条件下では流れを生み出すことによる (図 1.2)．例えば，イオンチャネル

9) ここでは説明を省くが，細胞の中の反応などによってチャネルを開ける確率は変わる．

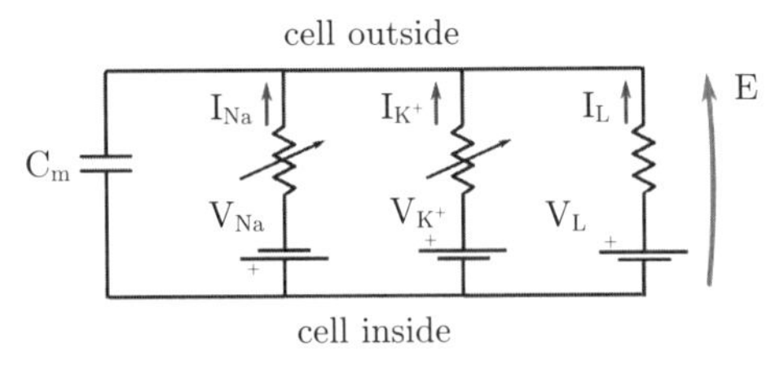

Fig. 1.3 The HH model equivalent circuit. The membrane is modeled as a capacitance C_m in parallel to generators. The three main currents are represented, I_L is due to external polarization and I_{Na} and I_{K^+} are generated by main ion pumps. Other minor ionic currents are neglected. The ion currents are variable and depend on the total membrane potential.

1.3 The action potential

How an action potential is generated was firstly clarified by Hodgkin and Huxley[Hodgkin and Huxley, 1952] in Cambridge, UK. They studied the giant neuron of the squid and developed a mathematical model which is able to reproduce almost exactly the electric behavior of the neuronal cell under different conditions. The model reduces the membrane to a simple electric circuit with few components: capacitance, resistors, and batteries.

The batteries in figure 1.3 represent the potential due to ion unbalance (mainly sodium E_N and potassium E_K). These two are variable depending on external conditions, whereas the E_l is a constant difference in potential given from outside. The two variable resistances R_N and R_K, and the constant R_l have a similar meaning. In this case the variable resistances depend on the membrane potential.

The model was derived using *patch clamp*[1)] experiments on the squid's giant axons, it shows that the properties of an excitable cell are described

1) An experimental method by which cells are kept at fixed potential through a hollow pipette connected to a circuit.

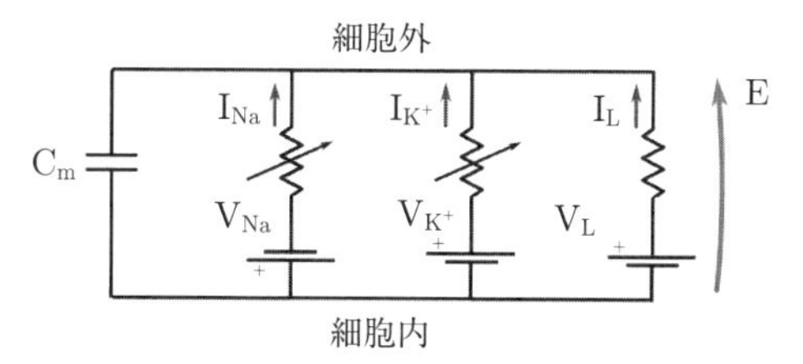

図 1.3　電気回路で示した H-H モデル．細胞膜は静電容量 C_m として並列に電源に接続されている．3 つのおもな電流が記されている．I_L は外部の分極によるもの，I_{Na} と I_{K+} はイオンポンプによるものである．その他の原因によるイオン電流は無視する．イオン電流は，全体の細胞膜のポテンシャルに応じて変化する．

は受容体（レセプター）としての機能ももつ．これは，あるチャネルの場所で特定な化学反応が起こると，チャネルが開く確率があがり，それにより電気的かつ化学的な事象の連鎖が導かれることを意味していて，結果として特徴的な電気信号であるスパイクや活動電位をもたらす．

1.3　活動電位

活動電位はイギリスのケンブリッジ大のホジキンとハクスレー [Hodgkin and Huxley, 1952] によって明らかにされた．彼らはイカの巨大軸索におけるニューロンについて研究を行い，異なる状況下における神経細胞の電気的な振る舞いをほぼ正確に再現することができる数学的なモデルを構築した．この H-H モデルは，細胞膜をコンデンサ，電気抵抗，そして電源を持つ電気回路として単純化したものである．

図 1.3 の中にある電源は，ナトリウムとカリウムの不均衡によるポテンシャルを表しており，それぞれをナトリウムの E_N とカリウムの E_K として記している．この 2 つの変数は，外部の状況に依存しており，E_l は外部とのポテンシャルの差から生じる定数である．2 つの変数 R_N，R_K と 1 つの定数 R_l も同様な意味をもつ．この場合，変数である抵抗値は，細胞膜のポテンシャルに依存する．

このモデルは，パッチクランプ法[10]を用いたイカの巨大軸索の実験から生

10) 細いガラス管を繋ぐことで細胞が一定のポテンシャルを保つことができる実験手法．

by these four differential equations:

$$I = C_m \frac{dE}{dt} + n^4 \frac{E - V_k}{R_k} + m^3 h \frac{E - V_{N_a}}{R_{N_a}} + n^4 \frac{E - V_l}{R_l} \tag{1.1}$$
$$\frac{dn}{dt} = \alpha_n (1 - n) - n\beta_n$$
$$\frac{dm}{dt} = \alpha_m (1 - m) - m\beta_m$$
$$\frac{dh}{dt} = \alpha_h (1 - h) - h\beta_h$$

This set of equations represents the three currents in the equivalent circuit for the cell's membrane in the sketch of figure 1.3. Notice that the voltage E is the membrane potential (in the circuit it will be the potential over the condenser C). What are the coefficients m, n and h? These are the population of open channels for each channel's type. Let's suppose that n is the population of potassium channels that are open. What is the number of channels dn that, after a small time dt, will close? Clearly, it will be the current number of channels n multiplied by the probability to close β_n. So we have that the variation of channels per unit time is $\frac{dn}{dt} \approx n\beta_n$. But there will be not only channels that close! In the interval of time dt, we will also have a population of closed channels that will open. If the population of open channels n is normalized to one, the closed channels will be $(1 - n)$. Defining α_n as the probability for a closed potassium channel to open, we will have that the increased number of open channels is $\alpha_n(1 - n)$. So, the total dynamics for the population of open potassium channels n is then the following:

$$\frac{dn}{dt} = \alpha_n (1 - n) - n\beta_n \tag{1.2}$$

Please notice again that the probabilities α or β are not constant, but depend on the membrane potential E. For example, in the original paper by Hodgkin and Huxley, α_n was

まれたもので，刺激性細胞が下記の 4 つの式で表すことができるというものである．

$$
\begin{aligned}
I &= C_m \frac{dE}{dt} + n^4 \frac{E - V_k}{R_k} + m^3 h \frac{E - V_{N_a}}{R_{N_a}} + n^4 \frac{E - V_l}{R_l} \qquad (1.1)\\
\frac{dn}{dt} &= \alpha_n(1-n) - n\beta_n \\
\frac{dm}{dt} &= \alpha_m(1-m) - m\beta_m \\
\frac{dh}{dt} &= \alpha_h(1-h) - h\beta_h
\end{aligned}
$$

この式は，図 1.3 に示す細胞膜と同等な回路に流れる 3 つの電流を表している．電圧 E が膜電位を表していることに注意してほしい．回路では，コンデンサー C にかかるポテンシャルともなっている．係数の m，n と h は何を意味しているのだろうか．これらは，それぞれのチャネルの種類ごとに，どのくらいのチャネルが開いているかを示している．ここで，n がカリウムのチャネルを開ける数だと考えてみよう．微少な時間 dt の後，どのくらいのチャネル dn が閉まるだろうか．もちろん，それはチャネルの数 n と閉じる確率 β_n をかけた値となる．したがって，単位時間あたりのチャネルの数の変化は $\frac{dn}{dt} \approx n\beta_n$ であることがわかる．しかし，チャネルは単に閉じるだけではない．微少な時間 dt のなかで，一度閉じたチャネルが再び開くこともある．チャネルが開く数 n を 1 と規格化した場合，閉じるチャネルの数は $(1-n)$ である．α_n を閉じたカリウムのチャネルが再び開く確率であるとすると，開くチャネルの数の増加は $\alpha_n(1-n)$ と記すことができる．そこで，カリウムのチャネルが開く数 n は，以下のように表すことができる．

$$\frac{dn}{dt} = \alpha_n(1-n) - n\beta_n \tag{1.2}$$

ここで，確率の α や β は一定ではなく，細胞膜のポテンシャル E に依存していることをもう一度思い出そう．例えば，ホジキンとハクスレーの論文では，α_n は

$$\alpha_n = a \frac{b - E}{e^{\frac{b-E}{c}} - 1} \tag{1.3}$$

$$\alpha_n = a\frac{b-E}{e^{\frac{b-E}{c}}-1} \tag{1.3}$$

where a, b and c are constants determined experimentally. Depending on $b-E$ sign, the trend of α will change and channel dynamic will invert, thus b is a sort of threshold value. If you repeat the same line of thinking for the other channels types, you will obtain the last three differential equations of the Hodgkin and Huxley model (see equation 1.1). Take into account that the two researchers (that got the Nobel Prize in 1963 for their work) determined experimentally α and β for all the channels and that the exponent of the three m, n and h populations (numbers between zero and one) were also determined in accordance with the experimental data. Now you know the type of reasoning used by Hodgkin and Huxley to model the electric activity of excitatory cells.

You may ask yourself, what kind of signal is the spike? Well, the inactive cell due to the continuous actions of ion pumps, stays at a constant membrane potential of about $-70\,\text{mV}$ (*resting potential*). Then, if some electrical changes occur, could be something happening at some receptor sites, or some electrical activity at the periphery (we will go in more details in later chapters), the sodium channels will be activated. In other words, the probability of closed sodium channels to open (α_m) will increase. More sodium channels open will make the cell's inside become less negative. If this phenomenon continues over a certain *gate threshold*, more sodium channels will open resulting in a steep rise of membrane potential. This phase is called *depolarization*. When another threshold is surpassed, the membrane potential inverts and becomes positive, the potassium gates begin to open, initiating the opposite phenomena (membrane *repolarization*). These channels pump ion faster than sodium, so the polarization process continues until an *hyperpolarization* stage occurs at about $-90\,\text{mV}$. In this phase, the

とされている．ここで，a，b，c はそれぞれ実験的に決定される定数である．$b-E$ に依存し，傾き α が変化し，チャネルの挙動は反転するが，b はその閾値（しきいち）の値である．他の種類のチャネルについて同様に考える場合には，ホジキンとハクスレーモデルから，式 1.1 の最後の 3 つの微分方程式を得ることができる．1963 年にこの研究によりノーベル賞を受賞した 2 人の研究者は α と β の値を，すべてのチャネルについて実験的に決定したが，m，n，h の指数についても，すべて実験のデータから決定した．ここまで読んであなたも，ホジキンとハクスレーのモデルを電気的な活性と，刺激性細胞に適応する理由がわかることでしょう．

さて，あなたはスパイクと呼ばれる信号について，疑問をもつことがあっただろうか．細胞は活動していないとき，イオンポンプは連続的に動くので，細胞膜のポテンシャルは -70 mV という一定値を保つ．これを**静止電位**[11]という．ここで，もしも何か電気的な変化が起こるとする．例えばどこかの受容体であるとか，そのほか電気的な活動が周辺で起こるとすると[12]，ナトリウムチャネルが活性化する．言い換えると，ナトリウムチャネルが開く確率 (α_m) が増えるということである．ナトリウムチャネルが開けば開くほど，細胞の内側の電位は上がる．もしもこの現象がゲートしきい値をこえるまで続いた場合，より多くのナトリウムチャネルが開くため，結果として細胞膜のポテンシャルは急激に上がる．このことを**脱分極**[13]と呼ぶ．細胞膜のポテンシャルが閾値を超えて，（つまり $b < E$ になると）α がプラスからマイナスへ転じる[14]．すると，競合するカリウムチャネルは開くという，先ほどとは逆の現象が起こり，細胞膜は再分極[15]する．これらのチャネルは，ナトリウムよりもイオンを早く排出するので，分極過程は続き，-90 mV で過分極が起こる．この段階において，カリウムとナトリウムのゲートは両方とも閉まる[16]．そして，細胞は 5 ms ほど中性（活性ではない状態）になり，カリウムとナトリ

11）静止電位：resting potential.
12）この周辺については後の章で述べる．
13）脱分極：depolarization.
14）α はつまり，ポテンシャルの上がり方（あるいは下がり方）を示している．
15）再分極：repolarization.
16）すべてが閉まるわけではなく，閉まったチャネルが開く確率が低くなるために，閉まっているチャネルが多くなる．

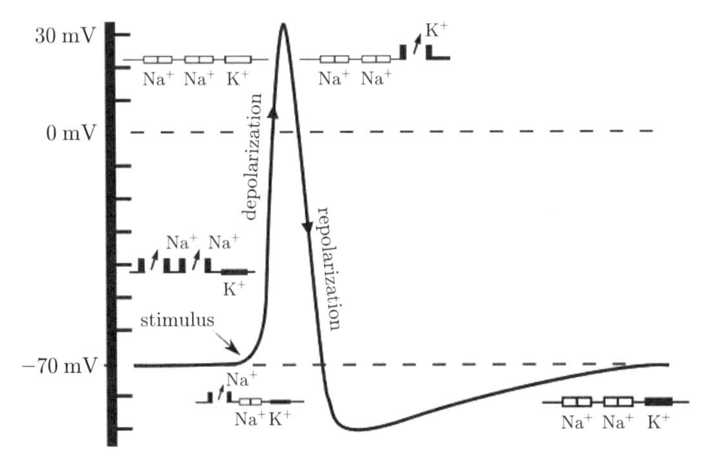

Fig. 1.4 A stimulus induces the sodium gates to open causing depolarization in the cell. As soon as the potential value rises, more sodium gates open in a positive feedback process. When the cell potential becomes positive the sodium gates begin to close and at about +30 mV the potassium opens. This provokes the opposite phenomenon of repolarization until the cell returns to the base −70 mV polarization. It takes about 5 msec for the cell to "recharge" and be able to spike again.

potassium and sodium gates are both closed[2] and the cells remain inert (it is not excitable) for about 5 msec. The activity of sodium and potassium pumps bring back the unbalance, the cell's membrane is again at its resting potential of −70 mV. Overall, all this depolarization-re-polarization process looks like a brief spike of electric potential, like in figure 1.4.

The inert period of 5 msec is not only a "recharging" moment for the cell. It is actually more important than it seems: it guarantees that signals proceed monotonically in one direction, and do not go backward. In fact, when a spike occurs and it is transmitted towards the next adjacent site inducing, say, another spike. This other spike will not induce back a spike on the first place because that is now inert. Without this property, we would have unstable neural networks, where signals would bounce back and forth in the same confined group of neurons.

2) They are not *all* closed, the probability for a closed channel to open is low, and the probability for an open channel to close is high.

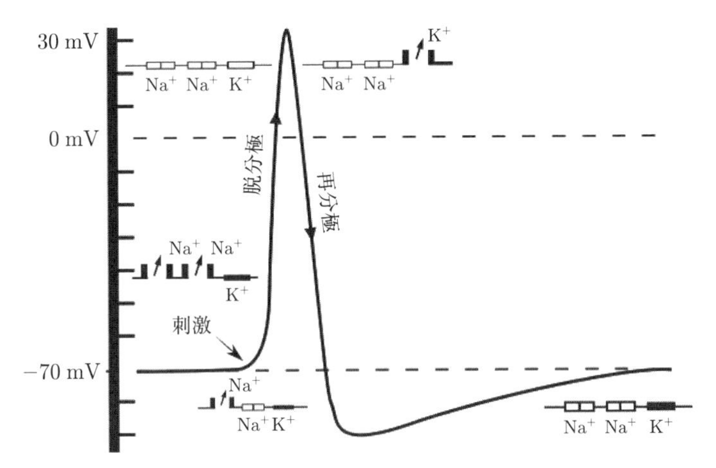

図 1.4　刺激があると，ナトリウムゲートが開き，細胞内に脱分極が起こる．ポテンシャルが上がるやいなや，より多くのナトリウムゲートは正のフィードバックを起こして開く．細胞のポテンシャルが 0 を超えて正になると，ナトリウムゲートは閉じ始め，ポテンシャルが +30 mV になると，カリウムのゲートが開く．そして最初と逆の現象が起こり，再分極の現象が，ポテンシャルが −70 mV に達するまで起こる．そして再びスパイクが可能になるまでは 5 ms の時間を必要とする．

ウムポンプの活性は，再び不均衡をもたらす．細胞膜は再び −70 mV の静止電位に戻る．全体にわたって，すべての脱分極と再分極の過程は，図 1.4 に示したように，電気的なポテンシャルの短時間のするどい信号のようである．

不活性な 5 ms の過程は，細胞にとって単に再充電する時間というわけではない．それよりも，信号は一方向にのみ伝わり，逆には伝わらないことを保証することのほうが重要である．事実，スパイクが起こるとき，それは近隣のサイトへ誘発を行なっていて，それはつまり次のスパイクを生む．この次のスパイクというのは，逆方向には誘発されない．なぜならそのとき，その場所はもう不活性だからである．この性質がないと，私たちは不安定な神経ネットワークをもつことになってしまい，信号は限られたニューロンの間を行ったり来たりすることになってしまう．

1.4 Taste

What are the senses? In very simple words, the so called five *senses* are just brain functions that detect particular physical process (light, temperature, acceleration, chemical composition or other) and processes it. One of the oldest sense is taste. We call it an *old* sense because, as far as we can understand, it is present in animals that lived millions of years ago. It is an essential sense that it is fundamental to give a first chemical assessment of food composition, giving the animal the ability to distinguish food from poison. Also, in a certain sense, taste is also present in archaic organism, like bacteria, where chemical composition of the media in which they are, can drive movement or particular responses.

In human, and in other higher mammal, the taste sense is located in the mouth on a specialized organ called *tongue*. On the surface of the tongue there are small pores (sort of holes called *taste buds*) where liquefied food is trapped. We should not confuse: taste deals only on the analysis of liquid matter, not solids. Any solid matter is firstly solubilized by *saliva*, a fluid produced by *salivary* glands in the mouth. The saliva is the solvent and the solute matter is analyzed. On the walls of the taste buds (see the sketch in figure 1.5) there are specialized receptor cells that are sensitive to the chemical composition of the salivary solution. When a certain molecule is captured by the receptor, the activity of the cell increases. In other words, the spike rate of the cell is augmented proportionally to the concentration of the chemical captured.

Science disagree on what are the basic tastes, generally there are five basic flavors that human can feel: sweet, bitter, sour, salty and umami. This means that there are receptor cells in the taste buds that react to different chemical in different ways. The response of the cell is coded into firing rate, for example in figure 1.6 we see experimental firing rates obtained exposing a rat tongue with increasing concentration of NaCl or

1.4　味覚

感覚とは何だろうか．簡単に説明すると，いわゆる五感とは，特定の物理的な過程（光，温度，加速，化学組成やその他）を検出し，それを処理する脳機能の一部である．最も古くからある感覚は味覚である．なぜ古くからある感覚かというと，数百万年前から動物に存在していることが解明されているからである．味覚とは，食べ物の材料の化学的な組成を最初に分析し，毒かどうかを判断するのに必要な本質的な感覚である．例えば海水の中にいるバクテリアなどの細菌は，水の中でどちらの方向に餌があり，どちらの方向に危険な物質があるのかを感知し，進む方向を決めることができる．

人間や，その他の高等哺乳類は，味覚は舌と呼ばれる特定の箇所にそのセンサーを有する．舌の表面には**味蕾**[17]と呼ばれる小さな穴があり，液化した食べ物を捉えるようになっている．味覚というのは固形物ではなくて，液体になっていないと感じないということを整理しておこう．どんな固形物も，最初に口中の唾液腺[18]によってつくられた**唾液**[19]によって液体に溶ける．唾液は物体を溶かして解析するための溶剤としてはたらく．図 1.5 にあるように，舌の味蕾の壁の上には，唾液に含まれた化学物質に敏感な，特別な受容体がある．ある分子がその受容体に捉えられると細胞が活性化する．言い換えると，細胞のスパイク率は，捉えた化学物質の密度に比例して増加する．

実は，科学的には味というものを厳密に定義することはできない[20]．人間は 5 つの基本的な味，甘味，苦味，酸味，塩味，そしてうま味を感じることができる．これは，味蕾の受容細胞はそれぞれ違う化学物質に対して，違う方法で反応しているということを意味する．細胞の反応は発火率にコード化されており，例えば，図 1.6 に示したように，発火率を NaCl またはスクロースの濃度を増やしながらネズミの舌に投下し，実験的に得ることができる．

---

17) 味蕾：taste buds.
18) 唾液腺：salivary glands.
19) 唾液：saliva.
20) 例えば，極端に甘いものを食べている地域で育った人は，他の地域のお菓子などは甘くないと思うし，逆も然り．つまり，糖度を数値として計測することは可能だが「このお菓子（果物）はどれだけ甘いです！」などと味を科学的には定義できないという意味．

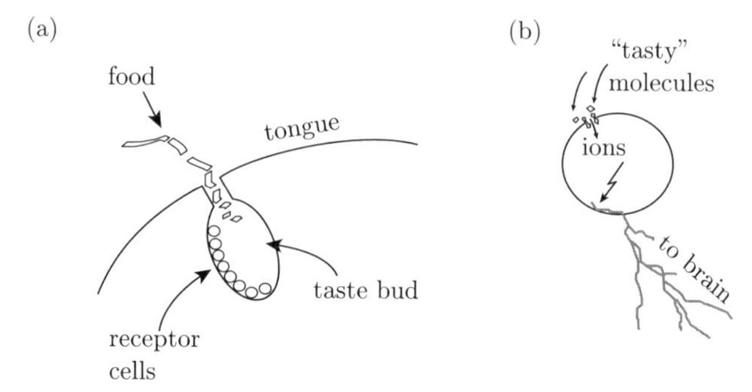

Fig. 1.5 The sketch of the tongue with taste buds. There are thousands of tastebuds on the surface of the tongue, and a sample is shown in (a) above. The taste bud is a small cavity in the tongue surface. Food is decomposed in different chemicals and captured by the bud. The internal surface of the taste bud is covered by receptor cells, schematically drawn in (b). On the cell membrane gates are activated by chemicals corresponding to a specific taste (for example bitter, salty, sweet etc.) When a gate opens, ions flow inside the cell and provoke depolarization and electric spikes that reach the brain.

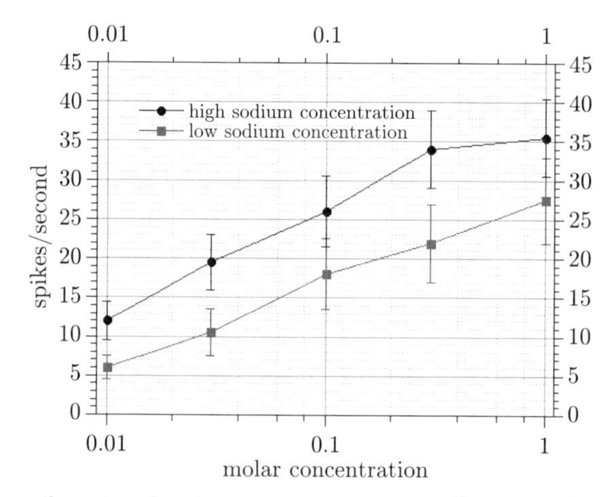

Fig. 1.6 The spike rate of rat's gustatory receptor cells exposed to variable concentration of salt (NaCl). Noticeably the spike rate is a predictor of the molar concentration of the chemical. The two curves represent two different conditions in which the measurements were taken[Tamura and Norgren, 1997].

sucrose[Tamura and Norgren, 1997].

The firing rates in the figure are not coming directly from a single taste bud's receptor cell. These signals are generally taken from single neurons

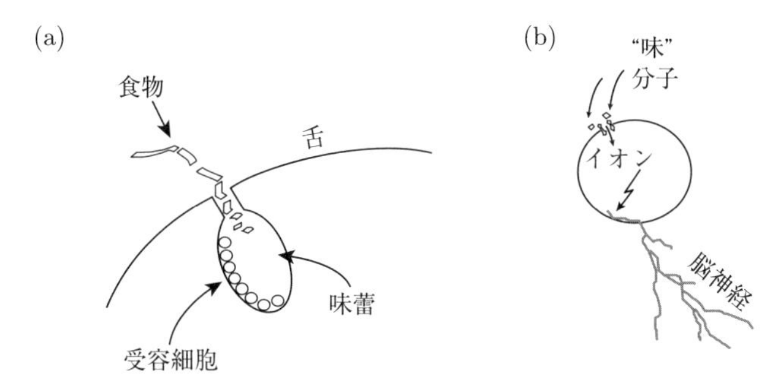

図 1.5 舌にある味蕾の模式図. (a) は舌の表面を示す. 舌の表面には何千という味蕾がある. 味蕾は舌の表面で小さな穴のようになっている. 食べ物は味蕾で化学種ごとに分解される. 味蕾の内部は受容細胞で覆われており, その受容細胞の模式図を (b) に示す. 細胞膜のゲートは, 特定の化学物質にのみ反応するようになっている. 例えば, 苦味, 塩味, 甘さなどである. ゲートが開くと, イオンは細胞の内部に流れ込む. それによって脱分極が起こり, 脳に電気的なスパイクとなって届く.

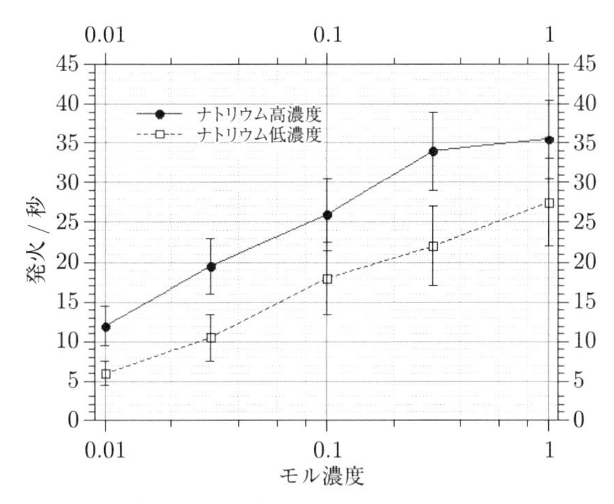

図 1.6 ラットの味覚の受容細胞が塩 (NaCl) の濃度に対してどのようなスパイク率の違いがあるかを示した図. 明らかにモル濃度が上がるほど, スパイク率は上昇する. 2つの結果は, それぞれ2つの違う状況において測定されたものである [Tamura and Norgren, 1997].

図の発火率は, 単一の味蕾の受容体からの結果ではない. これらの信号は, 一般的に複数の受容細胞の活性を集合させた単一ニューロンによるものである. 入力信号とスパイクがどのように関係しているのかということは (この場

Table. 1.1 Flavors thresholds in molar concentration

Taste	Molecule	Threshold for taste
Salty	NaCl	0.01 M
Sour	HCl	0.0009 M
Sweet	Sucrose	0.01 M
Bitter	Quinine	0.000008 M
Umami	Glutamate	0.0007 M

that aggregate the activity of many receptor cells. The way by which input signals, in this case the flavor molecule concentration and neural output, are correlated is called *neural coding* method. If a neuron responds with firing rates proportional to the molecular concentration, we say that the neural coding method is *rate coding*. Pure rate coding means that all the information about the stimuli is embedded in the neuronal firing rate (spikes per second).

This simple experimental evidence, is however not definitive and somehow controversial. There are numerous theories that suppose that information about stimuli and other things are coded in the brain with different and complex methods, see for example Brown[Brown *et al.*, 2004] for a review.

Humans can feel taste with different strength with good sensitivity. It is possible to measure a threshold of concentration of a certain flavor below which the taste is not felt at all, see table 1.1.

1.5 Propagation along the axon

Once a receptor is activated, a stream of action potential is generated and propagates up to the brain. As explained above, action potentials are *spikes* of electric signals, identical to each other. This train of signals has to reach the brain, that is tens of centimeters away from the tongue. In general, the axon of a receptor cell can be very long. If we think about tactile receptors in the lower regions of the human body, these axons can be even meters long!

表 1.1　味のモル濃度での閾値

味	分子	味の閾値
塩味	NaCl	0.01 M
酸味	HCl	0.0009 M
甘味	スクロース	0.01 M
苦味	キニーネ	0.000008 M
うま味	グルタミン酸	0.0007 M

合は味の分子の凝縮とニューロン信号の出力のことだが)，神経の符号化法と呼ばれるものである．もしも神経の発火反応が分子の濃度に比例しているとすると神経の符号化は発火頻度符号化として考えることとなる．純粋な発火頻度符号化は，刺激についてのすべての情報は，神経の発火率（1 秒あたりのスパイクの回数）に埋め込まれている，ということを意味している．

しかしながら，この単純な実験的根拠は最終的なものではなく，また，議論を呼んでいる．刺激に関する情報とその他の事柄は脳において複雑な方法でつながっていると仮説を立てる研究が多数ある．例えば Brown[Brown *et al.*, 2004] の参考文献を読むとよい．

人間は味を，それぞれ異なる強さで，敏感に感じることができる．特定の味ごとに，どのくらいのモル濃度になると感じるかという閾値を測定することが可能で，その結果を表 1.1 に示す．

1.5　軸索における伝搬

一度受容体が活性化されると，活動電位の素早い流れが引き起こされ，その後，脳へと伝搬する．前述のとおり，活動電位は電気信号のスパイクであり，スパイクの波形は常に同じである．この信号の連鎖は舌から数十センチのところにある脳まで届く必要がある．一般的に，受容細胞の軸索はとても長い．人体の下半身の触覚については，その軸索はなんと 1 m を超えるのである．

では，数 mV の一連のスパイクは，どのようにして，その電圧を変えずにそれほどの距離を伝うことができるのだろうか？　また，そのエネルギー源

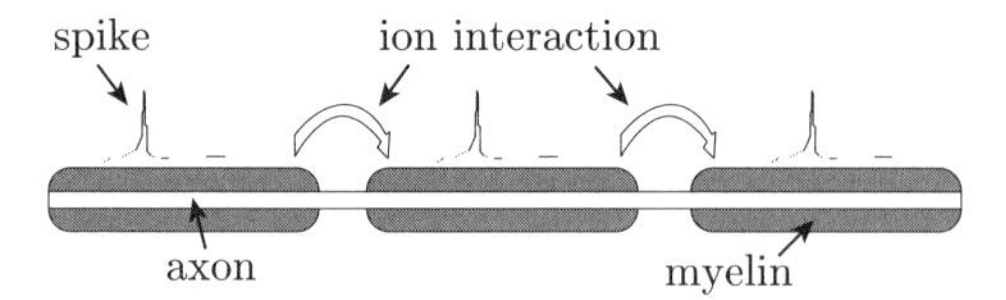

Fig. 1.7 The electric spike propagates between segments of the axon. The signal is kept isolated by *myelin* and propagates because of complex ionic interactions.

So how the few mV train of spikes can propagate so far away, without degradation? And what energy do they use? The same tricks used by the neuron's cell body apply to the axon. The axon is a cell filament that has the same ion content and membrane channels of the neuron cell body. So, simply speaking, the general mechanism of opening and closing of channels described above is valid. The key point to understand here is the role of scales (sizes of things in the cell). A cell is about few microns in diameter, but its axon length can reach easily meters, nearly millions of times larger. An axon can be thought as a tube of few microns of diameter, decimeters or meters long. With this in mind, we understand that the electrical state of the cell body, sometimes called *soma*, can be different from the one of the axon. Instead the electrical state changes along the axon. What has being said for the neural cell body, is valid for the axon locally. That is, in a confined region of the axon, the same mechanism of opening and closing of channels exists (see figure 1.7).

When the main body of the neuron *fires* ("spikes"), the depolarization propagates sideways, at a speed of about 100 meter per second, and also affects the channels on the part of the axon adjacent to the cell body. However, the action potential of the cell body soon reaches the hyper-polarization phase (sodium and potassium channels are closed), while somewhere else in the axon the electric state could be active, what happens in this case? The trick here is that the spike itself has a duration of about 1 msec only, whereas the hyperpolarization *locked* state is relatively long, about 5–10

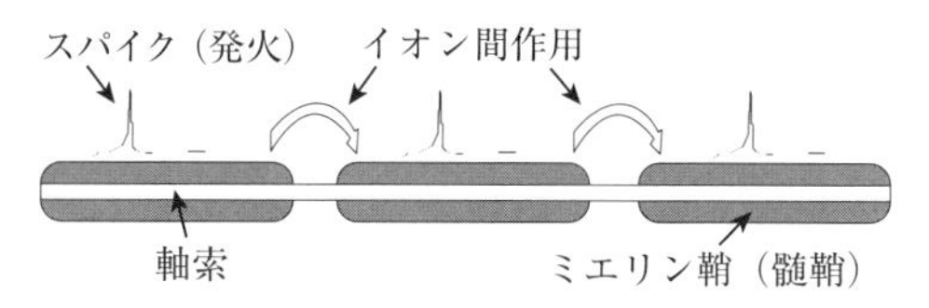

図 1.7 電気的なスパイクが軸索を通ってどのように伝達するかを示した図．電気信号は，髄鞘によって電気的に絶縁されている．この信号は複雑なイオンの相互作用によって伝達されている．

は何だろうか．実は，神経細胞体が用いる方法が軸索に適用されているのである．軸索は，細胞体と同じイオン及び膜チャネルを持つ線状の細胞である．つまり，簡単に言えば，前述したチャネルの開閉の一般的なメカニズムが当てはまるのである．ここで理解すべき要点は，細胞内器官のサイズの持つ役割であろう．細胞は直径数マイクロメートルというサイズであるが，その軸索は 1 m に届く場合もある．これは細胞の数百万倍もの長さである．軸索は直径数マイクロメートルの筒状の形状をしていて数十センチメートルから数メートルの長さをもつ．この点を念頭において，（神経）**細胞体**[21]と呼ばれる細胞の電気的な状態について理解しよう．細胞体は軸索とは電気状態が異なり，その電位は軸索に沿って変化したものである[22]．これは，これまでの神経細胞の電位の話（スパイク）は，局所的な軸索の電位の変化にも適応できることを意味していて，つまり，軸索と定義されている部分にもチャネルの開閉のメカニズムがあるということである（図 1.7）．

神経細胞（ニューロン）の発火，つまりスパイクが起こる時，脱分極が伝播する速さは 100 m/s ほどであり，軸索の（細胞に近い部分の）チャネルも影響を受けてスパイクが伝わってくる．しかし，細胞の活動電位はすぐに過分極の段階に達してナトリウムとカリウムのチャネルが閉まり，その間に軸索の電位は活性化できる．このとき何が起こるのだろうか．ここでのトリックは，スパイクそのものの持続時間は 1 ミリ秒であり，過分極は 5〜10 ミリ秒という長い時間，その状態を変えられないよう固定してしまうということである．この過分極の状態にあるチャネルはすべて閉じていて不活性である．その結果，

21）(神経）細胞体：soma.
22）電位が場所によって違うということを意味している．

msec. All the channel that are in that state are closed and inert. This results in the spike signal that propagates along the axon only in one direction (if the channels were not hyper-polarized and locked, they could be depolarized back by the activated axon gates, and no propagation would be possible). So the cell body triggers the action potential and on the axon it propagates one way only. The hyperpolarization state causes the shut down of the electric activity of the axon behind the spike signal.

From the energetic point of view, we have to consider that the axon stays in the resting potential of about $-70\,\text{mV}$ thanks to the work of ion pumps that consume ATP (adenosine triphosphate). ATP is a molecule that stores and transfers chemical energy within the cell. We do not have to study the chemical mechanisms of all these processes, that are actually very complex. In fact, there are other important details, for example the axon is covered in insulating matter, called *myelin sheaths*, where no ion exchange is possible. Ions are circulating only in the connecting points between these sheaths, called *Ranvier's nodes*. Thanks to the myelin, nervous signals are propagated faster and more efficiently. However, for our purpose of understanding how information is treated and communicated in biological signals, we only need to understand the big picture: the neuron cell body triggers an electrochemical chain reaction, called action potential, and this travels along the axon until it reaches the next neural cell. That's it for us!

1.6 Touch

Touch is a complex network of sensory apparatuses that reside in the human skin and on the skin of the majority of mammals and other animals. In humans these sensory apparatuses are differentiated in various forms. First of all let's consider the most evident touch sensor that we have: the *hair follicle*. Hairs are distributed on the skin in different density and lengths. These are long filament of keratin, a very tough and strong protein. Then, at the base of the hair is the follicle. It firmly locks the hair into the *derma*,

スパイク信号は軸索の一方向にのみ伝播する．仮に過分極にならず固定されていない場合，それらは，軸索のゲートを活性化させるように再び分極し，伝播は起こらない．つまり，細胞体は活動電位と軸索の一方向への伝播を引き起こす．過分極状態は，スパイク信号による軸索の電気的な活性を遮断するのである．

エネルギーという観点からみると，ATP を消費するイオンポンプのおかげで軸索は −70 mV の静止電位にとどまっている．ATP は細胞内で化学エネルギーを貯蔵し，伝達する分子である．この過程の化学的なメカニズムは実際とても複雑であり，勉強する必要はない．それよりも，例えば，軸索は髄鞘[23]と呼ばれるイオンの交換が不可能な絶縁体の物質で覆われているという事実の方が重要である．イオンは，それぞれの髄鞘の結合ポイントでのみ循環が可能であり，その結合ポイントのことをランビエ絞輪[24]という．髄鞘のおかげで，神経信号はより速く，そして効率的に伝播することができる．しかし，どのように情報が取り扱われ，生物学的な信号によって伝達されるのかということを知り，理解を深めるためには大きなイメージを理解すればよいだけである．それは，細胞体が活動電位と呼ばれる電気化学的な連鎖反応のきっかけを起こし，軸索を伝って次の神経細胞へ到達するというイメージである．そこが大切である．

1.6　触覚

触覚は，人間の皮膚に存在する感覚器官の複雑な回路網である．これは大多数の哺乳類やそのほかの動物の肌に備わっている感覚である．人間の場合，この触覚は様々な形態で区別されている．まずはじめに，最も明確なセンサーである毛包[25]について考えてみる．毛は異なる長さと密度で肌に分布している．

---

23) 髄鞘：myelin sheaths.
24) ランビエ絞輪：Ranvier's nodes.
25) 毛包：hair follicle.

the deepest layer of the skin. Hairs have many functions. One of the most important function in mammals is its contribution to the touch sense. In fact, hair keratin being strong and elongated, can detect vibration of the air and even certain sound frequencies. In which way does this happen? Well, the sounds move the air and the air push around the hair that, as a quasi-rigid body, transmits the movements down to the follicle. The touch sensor is not the follicle, but a long nerve filament that is wrapped around the base of the hair. If the hair moves, the nerve reacts with an electric spike, according to the same principles explained in the previous section.

What is this nerve made of and how does it produce a spike? You have to consider that this is a neuron cell with a very long specialized axon. If the nerve is deformed by a force, depolarization occurs. The biological mechanisms by which a mechanical deformation can cause the cell depolarization are not of importance to us. What we have to remember is that some process on the cell membrane is sensitive to deformation. If deformation occurs, ion channels begin to activate producing ion currents and cell depolarization. Everything happens in the same fashion as explained in the previous section, but this time the cause of the ion channels activation is not a chemical reaching the cell membrane, but mechanical: some force that deforms the cell membrane (see figure 1.8 for a sketch of the skin with the hair follicle and the dendrite cell wrapped around it).

The tactile sensation is unique, it is composed by multiple signals coming from different independent sensory systems. Besides the hair follicle, in the derma, there are small *tactile corpuscles*. These are clusters of elongated nucleus cells. These denser concentration of cells in the corpuscle create a confined region of higher mass density in the skin. If the skin vibrates the corpuscle resonates with a certain frequency that depends on the mass, shape and size of the corpuscle and its surrounding matter.

One type of this corpuscles is called *Meissner's corpuscle*, which is a round cluster of cells wrapped with nerves that will depolarize and become

毛はとても丈夫で強いタンパク質であるケラチンの長い繊維によりできている．そして，毛の根元が毛包である．この毛包によって，毛は肌の最も深い層である真皮にしっかりと固定されている．毛には様々な機能がある．哺乳類にとって最も重要な機能のうちの１つは，触覚への寄与である．実際，ケラチンは強くて長いので，空気中の振動や一定の周波数までも検出することができる．これはどのように起こるのだろうか．音は空気を振動させ，そして空気はとても硬い毛の周りを押し，その動きは下にある毛包へ伝わる．触覚センサーとは，毛包自体ではなく，毛の根元に包まれている長い神経線維を指す．毛が動くと，1.5 節で述べたのと同じ原理によってその神経は電気的スパイク反応を起こす．

この神経は何によってつくられ，そして，どのようにスパイクを起こすのだろうか．神経細胞は，とても長く特殊化された軸索，あるいは，軸索よりは短いが，長くなった細胞をもつことを念頭に置かなくてはならない．もしもこの細胞が力によって変形すると，脱分極が起こる．機械的な変形による細胞の脱分極の原因となる生物学的な過程は，ここでは重要ではなく細胞膜は変形に敏感であることを覚えておかねばならない．変形が起こると，イオンチャネルは活性化しはじめ，イオン電流が生じ，細胞は脱分極を起こす．その仕組みは，前に説明した内容と同じであるが，このとき，イオンチャネルの活性化の原因は化学物質が細胞膜に到達したことではなく，細胞膜に力が加わって変形したりするといった機械的なことに原因がある．図 1.8 は毛包を取り囲む触覚センサーを模式的に示したものである．

触覚の感覚はとても独特で，それぞれ異なって独立しているセンサーのシステムからくる多様な信号から成り立っている．毛包の他にも，真皮の上には**マイスナー小体**[26]というものがある．これは縦長の神経細胞の集まりである．小体の中に密集する細胞は肌のなかで密度が高い部分を引き起こす．肌が振動すると，小体及びその周辺の物体の質量，形，サイズに応じて小体は一定の周波数で共振する．

この小体の１つの種類がマイスナー小体と呼ばれるもので，神経に包まれ

26）マイスナー小体：Tactile corpuscles.

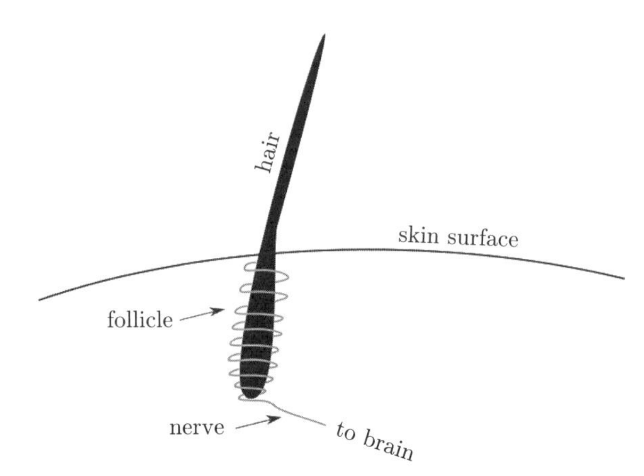

Fig. 1.8 The follicle touch sensor sketch. When the hair moves, it transmits its movement down to the nerve that it is wrapped around it. The vibration creates a stimulus that provokes a spike.

active when the skin is subject to vibrations of about 20–40 Hz. This is the way our skins feels low frequency vibrations. The *Pacinian corpuscles* are larger structures consisting of many concentric layers of cells. Inside this egg shaped structure there are nerve cells. If the corpuscle is deformed by pressure, the nerve cells react with action potential activation. Also, this corpuscle acts as a frequency sensor. Differently from the Meissner's one, the Pacinian corpuscle responds to higher frequencies in the range of 150–300 Hz. How does a corpuscle inside the skin tissue operate as a frequency detector? We have to consider that the corpuscle has different density than the surrounding skin. The difference in density works as a mass suspended in an elastic media. Oversimplifying, the mechanical system is like a mass attached to a spring. The spring represents the equivalent elasticity of the tissue in which the corpuscle is immersed.

The mass (the corpuscle) will then vibrate according to the simple physics of a mass attached to a spring. The ideal equations for this system is as the following:

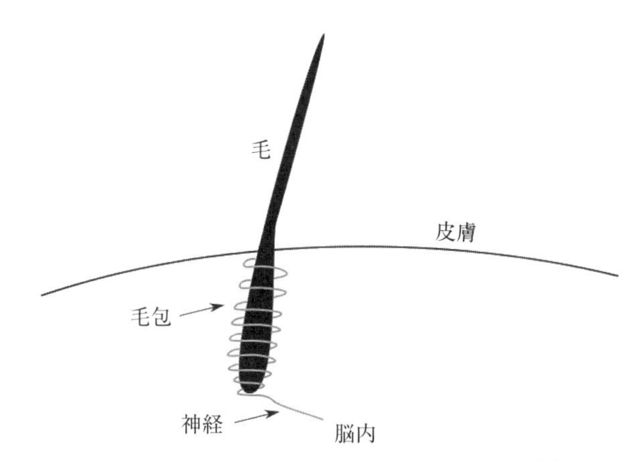

図 1.8 毛包と触覚センサーの模式図．毛が動くと，その動きは毛包を取り囲んでいる神経へと伝わる．この振動が刺激となり，スパイクが起こる．

た細胞の集合で，20〜40 Hz の振動を感じたときに脱分極を起こし活性化する．これが肌が低周波振動を感じる仕組みである．それに比べてパチニ小体[27]は大きな構造をしており，同心円状の層からなる．卵型をしたこの構造の内側には，神経細胞がある．この小体が圧力によって変形すると，神経細胞は活性化ポテンシャルとともに反応を起こす．また，この小体は振動センサーとしてもはたらく．ただ，マイスナー小体と異なり，パチニ小体は 150〜300 Hz という高い周波数域で反応する．では，肌組織の内側にある小体はどのように周波数センサーとしてはたらくのだろうか．小体は肌組織とは異なる密度をもっている．そのため，この密度の違いがばねのような弾性体につり下げられた質点のようにはたらくのである．単純化すると，ばねに繋がった質点のような系となっている．ばねは，小体が存在している肌組織に均一な弾性力を表している．

質点（つまり小体のこと）は極めて単純な物理の法則に従って振動する．理想的な状況の場合，ばねの振動は次のようになる．

27) パチニ小体：Pacinian's corpuscles.

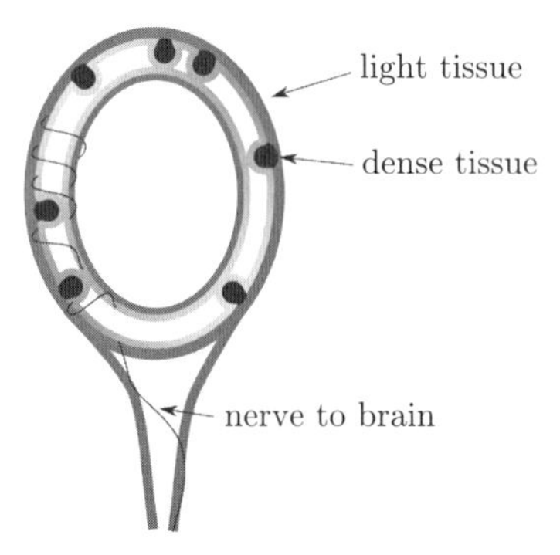

Fig. 1.9 A sketch of the Pacinian corpuscle in the skin. This functions like a mass-spring system where a lighter elastic tissue surrounds smaller denser tissue (indicated by the small dark circles in the picture). When a mechanical stimulus is applied, the skin vibrates and the nerves transmit the spiking up to the brain.

$$F = -K_{eq} * x \tag{1.4}$$

$$m * a + K_{eq} * x = 0$$

$$m\frac{d^2x}{dt^2} + K_{eq} * x = 0$$

This is the differential equation of an oscillatory motion of resonant frequency $\omega = \sqrt{\frac{K_{eq}}{m}}$, where K_{eq} is the elastic constant on the tissue used in the *Hooke's* law and m is the mass of the corpuscle (see figure 1.9). You have to consider that, in this simplification, the mass and the elastic constant are not directly definable. They are just abstract values that depend on the difference in density between the skin and the corpuscle and the elastic constant of the skin tissue. This simple model, however, gives you an idea of the basic mechanisms involved.

With this trick, the human body is able to calibrate different tactile vibration sensors, changing its sensitivity range by modifying the dimensions and mass of the corpuscle and the thickness and elasticity of the tissue where it is embedded.

Similar to the Pacinian corpuscle, another round structure of skin tissue

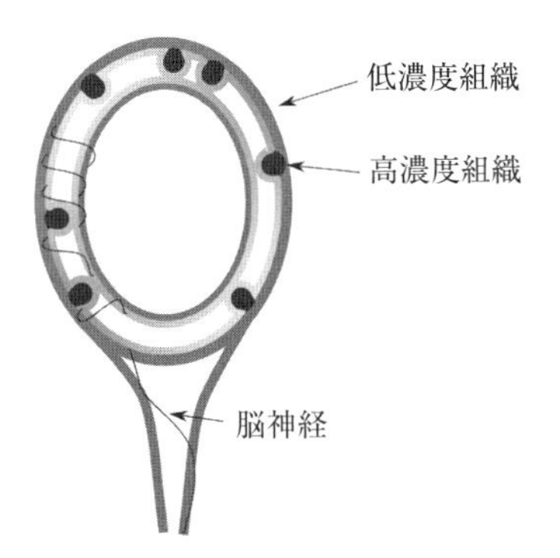

図 1.9 肌の中のパチニ小体の模式図．この機能は，ばねにつながれた質点（おもり）の系と同様である．つまり図の黒い部分が質点（おもり）で，周りの組織は弾性をもつので，ばねにつながれた質点（おもり）としての運動をする．機械的な刺激があると，肌は振動し，神経は脳にスパイクを伝える．

$$F = -K_{eq} \times x \tag{1.4}$$

$$m \times a + K_{eq} \times x = 0$$

$$m\frac{d^2x}{dt^2} + K_{eq} \times x = 0$$

これは共振周波数 $\omega = \sqrt{\frac{K_{eq}}{m}}$ における振動の微分方程式であり，K_{eq} はフックの法則[28]で用いられる肌組織のばね定数，m は小体の質量である（図1.9）．このばねと質点を用いた単純化のとき，質量とばね定数は具体的に何かを定義付けることができない．これらは抽象的な値であり，肌のなかの小体の密度の違いや肌組織の弾性力の違いに応じて変化するものである．これは単純なモデルだが，基本的な仕組みの考え方を示している．

小体の大きさと質量と，その小体が入っている皮膚組織の厚さと弾性力を調整して感度範囲を変えることで，人間は異なる触覚振動センサーを測定することができる．パチニ小体と同様に，皮膚組織による神経とともに丸い構造で存在するクラウゼ小体[29]と呼ばれるものもある．この構造は，圧力と冷たさに反応する構造をもっている．

28) フックの法則：Hooke's law
29) クラウゼ小体：Krause corpuscle.

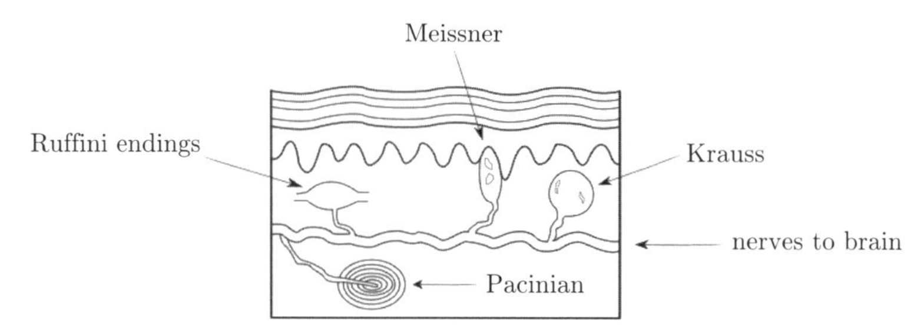

Fig. 1.10 The skin is a complex system of sensors that detect vibrations, temperature and chemical substances on the skin. Few are represented schematically in this picture. The spikes from these specialized cells are brought to the brain by nerves.

with nerves located inside it is called the *Krause corpuscle* or *Krause end-bulbs*. This structure responds to pressure and also is known to respond to cold.

Near the surface of the skin there are *Ruffini endings*. With this name are called several types of free nerve endings that respond to mechanical stimuli, temperature and even chemical stimulation. So, it should be clear to the readers, what we call the sense of *touch* is actually a very complex set of different sensors realized by diverse structures of tissue and nerves in the skin (see figure 1.10). The brain is able to integrate all these various mechanical and thermal stimulation signals, and responds with particular feeling of pressure, vibration, warmth, cold or pain. A more general word for this set of sensorial apparatus is *somatosensory receptors* system.

1.7 Olfactory

The olfactory sense is one of the oldest sensing apparatus in mammals. The organ specialized for this is the *nose*. In taste, liquids are chemically analyzed by the taste buds. In the same way, the olfactory sense does similar chemical analysis on gases circulating in the nose. The *inner nose* (nasal cavity) is the region inside human nose that does this chemical analysis. The inside of the nose is covered by a yellow pigmented tissue that contains

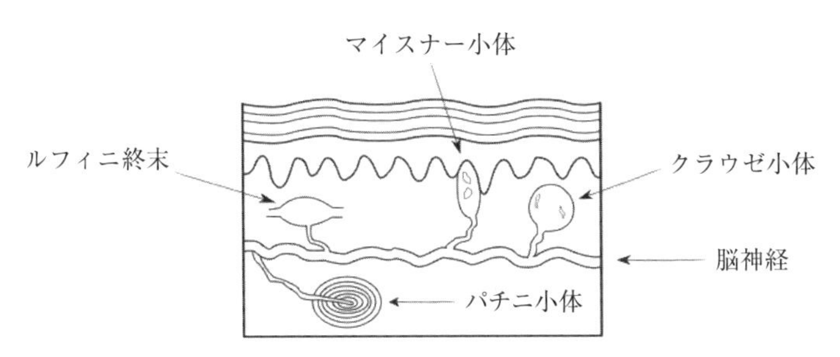

図 1.10 肌は，振動，温度，化学物質を検知する複雑なセンサーシステムである．図にこれらを模式的に示す．それぞれに特殊化された細胞からのスパイクは，神経を通り脳へと伝えられる．

皮膚の表面にはルフィニ終末[30]がある．機械的な刺激，温度や化学的な刺激に反応する数種類の自由神経終末[31]は，この名前で呼ばれる．

つまり，私たちが触覚と呼ぶ感覚は，皮膚組織の多様な構造と，神経を通って伝えられる違う種類のセンサーの実に複雑な集合であることがわかる（図 1.10）．脳は様々な機械的・温度による刺激の信号を統合し，特定の圧力による感覚や振動，暖かさ，冷たさ，または痛みとして反応することができる．これらの感覚器官を総称して，**体性感覚受容体**[32]と呼ぶ．

1.7 嗅覚

嗅覚は哺乳類における最も古い感覚の器官の 1 つである．そして，この感覚に特化した器官が鼻である．嗅覚の仕組みは味覚に似ている．味覚の場合は味蕾によって液体が化学的に分析されるのと同じように，嗅覚の場合には，

30) ルフィニ終末：Ruffini endings.
31) 自由神経終末：free nerve ending.
32) 体性感覚受容体：somatosensory receptors.

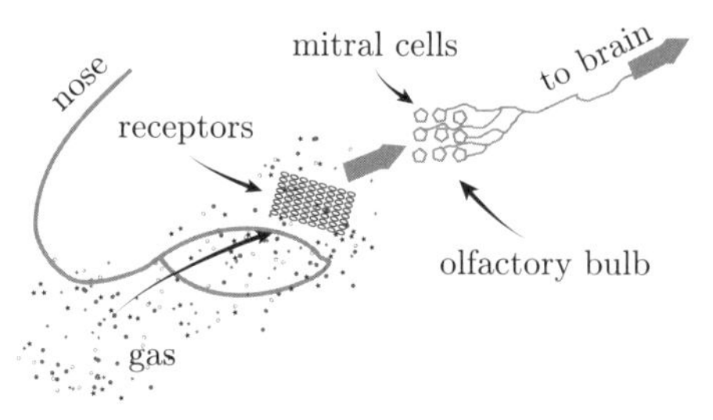

Fig. 1.11 A sketch of the nose and its neuronal system: when the gas enters the inner nose, thousands of chemoreceptors react and send the signal to the *olfactory bulb* area where *mitral* cells elaborate the signal and send it to the brain through nerves.

chemoreceptors, specialized olfactory neurons that respond to certain gas molecules (see figure 1.11). This receptive area covers a surface of about $2.5\,\text{cm}^2$ in total with about 25000 receptor cells. Signals from these cells are aggregated in the *olfactory bulb* region where *mitral cells* are present. There, signals are somehow elaborated and sent over to the brain by nerve fibers.

You should think about chemoreceptor as the same receptors we studied above about the taste. They function with the same principles: a molecule binds to the cell external membrane where the receptor cell induces a chemical mechanism that activate ion gates. These ion gates produce flux of ions that depolarize the nerve cell body and induce a chain of electrochemical events that generates a *spike* of electrical activity. Even if the chemical mechanisms and physical processes that induce the spike are different for every sense (taste, touch, smell, etc.) the shape and amplitude of action potentials are identical to each other. This demonstrates that the nervous system acts as a general communication network, which is able to deal uniformly with information coming from completely different organs, which is very interesting.

It seems that humans can distinguish 6 fundamental olfactory sensations: fruity, flowery, resinous, spicy, foul (rotten) and burned. However, these are complex feelings that can vary from person to persons and among cul-

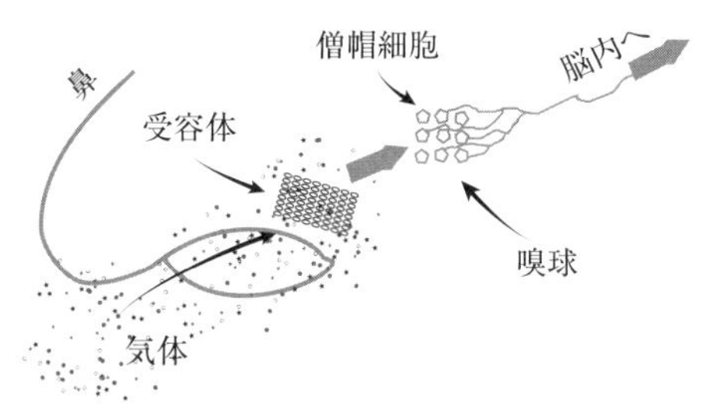

図 1.11　鼻と神経伝達の模式図．鼻の内部に気体分子が到達すると，何千個もの化学受容体が反応し，その信号を，僧帽細胞が情報を精査する嗅球部分に伝達したのち，神経を通って脳へと伝わる．

気体を鼻で化学的に分析する．**鼻腔**[33]は人間の鼻の内側の領域をいい，化学分析を行うところである．鼻の内側は黄色い色素をもつ化学受容体をふくむ組織に覆われていて，特定の気体分子に反応するように特化されている（図 1.11）．この受容体がある領域では，2.5 cm^2 に 25000 個もの受容細胞で覆われている．これらの細胞からの信号は**僧帽細胞**[34]が存在する**嗅球**[35]に集められる．そこでは，何らかの方法で信号が精巧につくられ，神経回路を通って脳へと送られる．

化学受容体は前に味覚について学んだ受容体と同じものと考えればよい．たしかに，化学受容体も同様の原理で，分子は受容体の外部の細胞膜と結合すると，細胞膜にある受容体が反応し，イオンゲートを開く化学的な仕掛けがはたらく．これらのイオンゲートはイオンの流れを生み出し，神経細胞の脱分極を起こす．そして，電気的に活性なスパイクを起こす一連の電気化学的な連鎖を生み出す．たとえスパイクを起こす化学的な仕組みや物理的な過程が味覚，触覚，嗅覚などそれぞれの感覚ごとに異なるとしても，その活動電位の形や振幅は同じである．これはとても興味深く，神経系がまったく異なる器官からきた情報を均一に処理できる一般的な通信ネットワークであることを示している．

人間は 6 つの基本的な匂いを嗅ぎ分けることができる．果物の香り，花の香り，樹脂の香り，香辛料の香り，腐敗の匂い，そして焦げた匂いである．し

33）鼻腔：inner nose または nasal cavity など．
34）僧帽細胞：mitral cell.
35）嗅球：olfactory bulb.

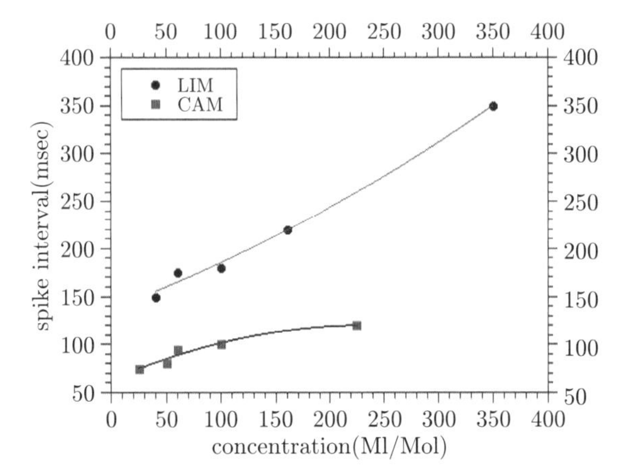

Fig. 1.12 Example data on how substances (LIM is derived from lemon, CAM from camphor) induce different spike rates on a Mitral cell. The spike rate is measured here as *interspike interval*. The smaller the interval between spikes, the higher is the frequency.

tures. The actual physical process is a mere electrical discharge of action potential (spikes) from neurons located in the nose. Again, this neuronal system responds with random spikes more frequently as the concentration of the target molecules becomes higher. In figure 1.12 we see an example in which the average spike frequency of different substances is compared with the molecular concentration. Clearly, different substances have different baselines, inclination and characteristics.

The olfactory system is much more sensitive than the taste system. In molecular concentration scale, the smell is about 10000 times more sensitive than taste. Actually, the sensation of taste is the result of the interplay between taste system and the olfactory system. The olfactory plays the major role in creating the final taste of food. For example, the good flavor of chocolate is caused mainly by its smell, the taste it is secondary. That's why food feels almost tasteless and less enjoyable when we have a cold and our nose cannot work properly.

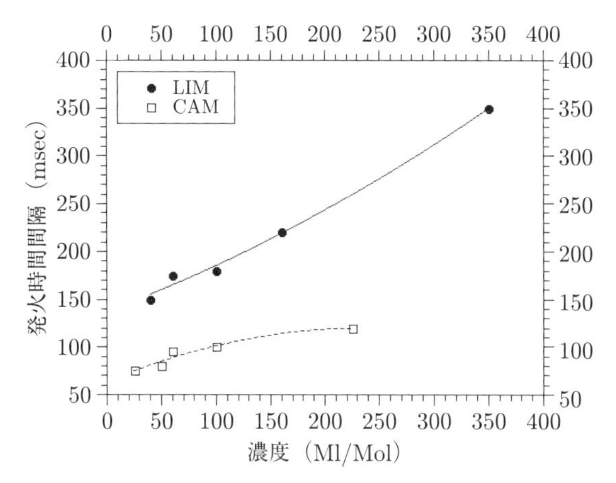

図 1.12　僧帽細胞におけるレモン (LIM) と樟脳 (CAM) の濃度とスパイク率のデータ．スパイク率は，スパイク間の時間（発火時間間隔[36]）によって測定された．発火時間間隔が小さいということは，頻度が高いことを示している．

かしながら，この感じ方は人によっても異なり，文化によっても異なるというとても複雑なものである．実際の物理的な過程は，鼻の中の神経細胞からの活動電位（スパイク）のわずかな電荷の放出である．繰り返しになるが，対象となる分子の濃度が高くなるほど，そのスパイクが起こる頻度は高くなる．図 1.12 に様々な種類の物質の濃度による平均のスパイク頻度のグラフを示す．物質によって基準線，傾向，性質が異なることは明らかである．

嗅覚というのは味覚よりもはるかに繊細である．分子の濃度の度合いを考えると，嗅覚は味覚の 1 万倍もの敏感さをもっている．そして，味覚というのは，味覚のシステムと嗅覚のシステムの相互作用によるものである．嗅覚は食べ物の味を最終的に判断するのに重要な役割を演じる．例えば，チョコレートの良い香りは，匂いがおもな要因であり，味は二次的なものである．これが，風邪などをひいて鼻が利かなくなったとき，多くの食べ物は味気がなくなり，その味を楽しめない原因である．

36) 発火時間間隔：interspike interval

1.8 Hearing

As you can imagine, hearing is a very important sense. Even when you read this text, you are actually *hearing* it in your mind. You recall the sound of the words being read (!) Well, how does the hearing work? Hearing consists of a system to capture and analyze sounds, but what are sounds? Sounds are pressure waves, and the vibration of air pressure is somehow detected by our special organ: the *ears.*

Then the next question is: how does an ear work? Centuries of anatomical studies show that the ear is really a marvelous device. Our ears have a protrusion called *pinna*, that is what we commonly call *ear.* In the center of the pinna there is an aperture, the entrance of the *ear canal.* This canal guides sound waves to the *ear drum.* The ear drum is a flat bony membrane that blocks the ear canal. On the other side of the ear drum the canal continues and reach outside, arriving up the nose (this part of the canal is called the Eustachian tube). This is an important fact: the pressure on the ear drum is acting from two sides (the nose side and the ear side), so violent sounds or sudden pressure variations have less impact on the ear, because forces coming from both sides will compensate.

When a sound arrives, its pressure waves stimulate the eardrum to slightly vibrate. As said above, the eardrum is a large round surface that completely blocks the ear canal, but also there are a system of three little bones (the *malleus*, *incus* and the *stapes*), connected to it. These bones function is to transmit the drum vibrations to another organ: the *cochlea*, a part of the so called *inner ear.*

You should imagine the cochlea as a tissue sack filled with a watery liquid called *perilymph.* The vibrations picked up by the ear drum are transmitted by those three small bones and exert a force on the liquid that vibrates as well, in tune with the sounds. The cochlea contains a canal (the *auditory canal*) with specialized cells that are able to respond to specific vibrations

1.8　聴覚

あなたのご想像通り，聴覚はとても重要な感覚である．この文章を読んでいるときでさえ，実際に頭の中で聞いているのではないだろうか．言い換えると，読んでいる言葉の音を思い出しているのである[37]．では，聴覚はどのように機能しているのだろうか．聴覚とは，音を捉えて解析する組織であるが，音とは何だろうか．音とは圧力波（疎密波）であり，空気の圧力の振動が，私たちの特別な器官である耳によって検出されるのである．

次の質問は，耳はどのようにはたらいているのだろうか？　何百年におよぶ解剖学的研究の結果，耳とは実際ものすごい装置であることがわかっている．耳には**耳介**[38]と呼ばれる突起があり，一般的にこれを耳と呼んでいる．耳介の中央には穴があり，それが**外耳道**[39]の入り口である．この穴を通って音は**鼓膜**[40]に届く．鼓膜は平らな骨のような膜であり，外耳道をふさぐものである．鼓膜の反対側は外耳道が続いており，外側へと達し，鼻と通じている（この外耳道の部分は**耳管**[41]と呼ばれる）．鼓膜での圧力は 2 つの側（鼻側と耳側）の両方につながっているという，この事実は重要である．つまり，ものすごい大きな音や，急な圧力の変化があった場合，鼻の側からもその圧力が到達することで，相殺できる仕組みになっており，耳への衝撃を減らすことができるからである．

音が届いたとき，その圧力は鼓膜を刺激し，少し振動する．前述のとおり，この鼓膜は大きな丸い表面をもっていて外耳道を完全にふさいでいるが，**槌骨**[42]，**砧骨**[43]，**鐙骨**[44]という 3 つの小さな骨が鼓膜につながっている．これらの骨の機能とは，鼓膜の振動をもう一方の器官である**内耳**[45]の一部である

37) 例えば，手紙を読むときに，送ってくれた相手の声が聞こえるといったことがあるだろう．
38) 耳介：pinna.
39) 外耳道：ear canal.
40) 鼓膜：ear drum.
41) 耳管：eustachian tube.
42) 槌骨（つちこつ）：malleus.
43) 砧骨（きぬたこつ）：incus.
44) 鐙骨 (あぶみこつ)：stapes.
45) 内耳（ないじ）：inner ear.

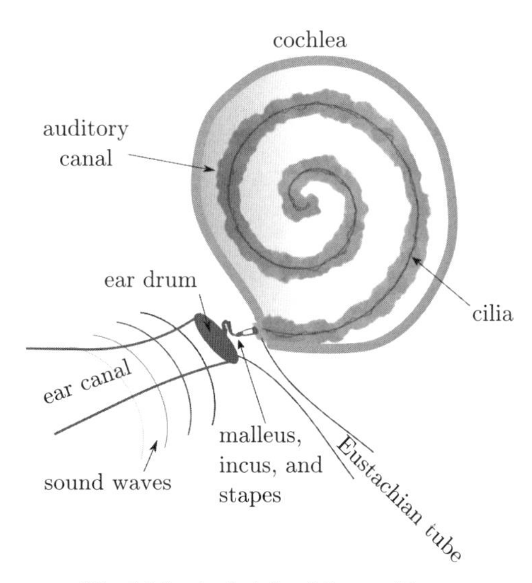

Fig. 1.13 A sketch of the cochlea.

(see figure 1.13). These cells have microscopic hairs on its top (*cilia*), the hairs produce vibrations and the cells respond in the same way tactile cells described in the previous section do: vibrations induce cell deformation and the specialized channels on the cell's membrane become active and induce the spike. Same principle as above. So, we now understand that the perilymph liquid vibration can induce neural response, but how can this neural response be in tune with the vibration frequencies? Well, it isn't! We can understand that by the structure of the cochlea: the shape of the cochlea is a spiral and so it is the auditory canal. So sounds, depending on their wavelength, will induce vibrations of different intensity along the spiral. Deeper and deeper in the canal vibration receptors are allocated, all similar to each other.

How does this work? Please notice that the physical wavelength of audible sounds is not of great help to distinguish sounds for humans. In fact, the sound wavelength can be calculated by the following:

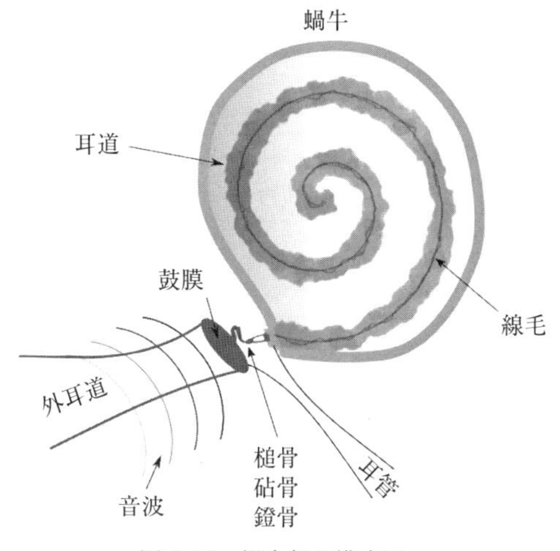

図 1.13 蝸牛殻の模式図.

蝸牛殻[46]に伝えるためのものである.

蝸牛殻とは水のような液体である外リンパ[47]で満たされている袋のような組織である. 振動は鼓膜によって拾い上げられ, 前述の3つの小さな骨に伝搬し, この液体へ力を加える. こうして音を伝えるのである. 蝸牛殻は耳道[48]と呼ばれる管をもっており, そこには特定の振動に反応することができる細胞がある (図 1.13). この細胞はとても小さな線毛[49]が生えており, この毛が振動を生み出し, 細胞は前述した触覚細胞と同じ方法で, その振動を感知する. 振動は細胞を変形させ, 特定の細胞膜のチャネルを活性化させ, そして脱分極を生み出し, スパイクを起こす. 他の感覚と同様である. さて, 外リンパ液の振動が神経細胞を反応させることを理解したが, 神経細胞はどのようにして振動の周波数に反応できるのだろうか. 実は, 細胞が周波数に反応しているのではない. 蝸牛殻の構造はらせん型になっていて, らせん状の部分に届く音は波長ごとに異なる. 深くなればなるほど, 振動を受け取る受容体が割り当

46) 蝸牛殻 (かぎゅうかく) : cochlea.
47) 外リンパ : perilymph.
48) 耳道 : auditory canal.
49) 線毛 : cilia. 繊毛ともいう.

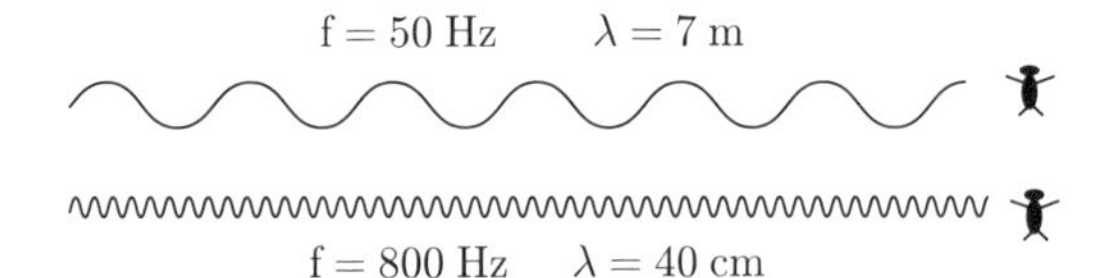

Fig. 1.14 Since sound frequency can range from few up to several thousand Hertz, the wavelength can variate a lot. The figure above shows two frequencies of 50 Hz (top) and 800 Hz (bottom) respectively, with the silhouette of a human (note that wave sizes are not in scale for clarity). Low frequencies have much bigger wavelength than the human body, others can be much smaller. This fact has an important influence on how sounds propagate and how we hear sounds.

$$\text{wavelength} = \frac{\text{propagation speed}}{\text{frequency}} \tag{1.5}$$

The sound speed in air, $v_s = 345$ meters per second, divided by the frequency of vibration gives us the amount. So for very low sounds, down to fifty Hz for example, we have very long waves, of more than 5 meters. For sounds in the range of human speech frequencies (100–300 Hz) we have wavelengths of one to three meters and even higher frequencies to the limits of audible (for example 15000 Hz) correspond to about 1 centimeter wavelength (see figure 1.14). But the human cochlea is less than one centimeter in total, so the size of the cochlea and the ear cannot be of any help to discriminate frequencies (note that sound waves are even larger in liquids, because sound speed in liquid become faster: about 1500 meters per seconds). So how do we hear: how can the cochlea distinguish different frequencies? The trick is that the cochlea is a tube divided in half along its length by a membrane that has different stiffness. So a sound stimulate vibration only in the place where the stiffness is best to resonate. In other words, the cochlea behaves like an array of mechanically resonant filters with detectors (the cilia)[Ashmore and Geleoc, 1999]. Cool enough!

As any other receptor cell, the output spike train is guided to the brain by long axons, or nerves. The brain is able to recognize the frequency by simply discriminating from what position in the cochlea the signal is coming from. So the brain is actually using a *spatio-temporal* approach to give meaning

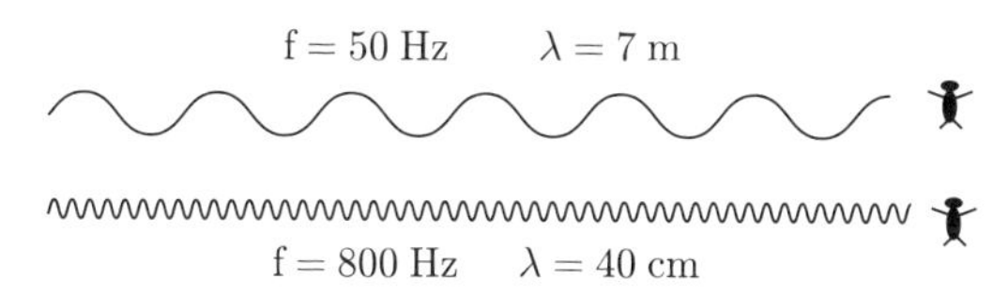

図 1.14　人間が聞くことのできる音の振動数は数 Hz から数千 Hz まであるので，波長もその範囲で変化する．図は 50 Hz と 800 Hz の波を，右に示した人間のシルエットとともに，そのスケールを示したものである．周波数が小さいと，波長は人間の体よりも大きくなり，周波数が大きいと，波長は人間に比べて小さくなる．これは，音がどのように伝播し，私たちがそれをどのようにして聞いているのかということに重要な影響を及ぼしている．

てられており，それらは互いに似ている．

蝸牛殻はなぜ音を聞くために，このような複雑な形をしているのだろうか．ここでは，音の波長というものは音を聞き分けるのに重要なものではないことに注意しよう．音の波長は，次の式で求められる．

$$波長 = \frac{伝搬速度}{振動数} \tag{1.5}$$

ここで，空気中の音の速さは $v_s = 345$ m/s であり，波長はこれを振動数で割ったものである．つまり，とても低い音，例えば 50 Hz などの場合，波長はとても長くなり，5 m を超える．人間が喋る領域の 100〜300 Hz の場合，波長は 1〜3 m 程度であり，もっと高い周波数帯，例えば聞こえる限界の 15000 Hz では波長は 1 cm となる（図 1.14）．しかし，人間の蝸牛殻は全部で 1 cm もない．つまり，蝸牛殻や耳の大きさは，周波数を識別するのに何の助けにもならない（ただし，液体中の音波の場合，その速さは空気中よりも速くなり，約 1500 m/s にもなるため，大きくなることに注意する）．では，私たちはどのようにして音を聞いて，そして，違う周波数を認識できるのだろうか．そのからくりとは，蝸牛殻は，違う硬さをもつ膜によって管の太さが半分に分けられていることにある．つまり，音は，ある特定の一番よく共鳴する振動数のときに最も刺激が起こる．違う言葉で言い換えると，蝸牛殻は機械的な共振フィルターと探知機（線毛）の配列のようなはたらきをする [Ashmore and Geleoc, 1999].

他の受容細胞と同様に，出力スパイクの連鎖は長い軸索や神経を通って脳へと伝わる．脳は蝸牛殻のどの位置からきた信号かを認識することができる．つ

to the spike trains, we will come back to this later.

Inside the same watery liquid sack, there is not only the spiral organ called cochlea that we just describe, there are also three ring of tissue. These rings are filled with the same perilymph liquid and are placed at the angle of 90 degrees to each other, like the three orthogonal axes XYZ. These are called *semicircular canals.* In Humans, every time our head moves, the liquid in the canal stimulates the cilia that produce vibrations and consequently neural spikes in similar way as in the cochlea. However, the brain does not interpret them as sounds, but as movements of the head. Because there are three of these semicircular canal, and because they are oriented orthogonally to each other, the brain can discern in what exact direction in space the head is moving. These knowledge creates the sense of *equilibrium.* This sense has nothing to do with the hearing, but its sensory apparatus is located very near the ear, a fantastic efficient way mother Nature utilizes space and resources!

To understand better about how the brain elaborates sounds, it is worth to mention about the human *echolocation* abilities. This particular function is very interesting. The question is "how does the brain recognize the direction of sounds?". The quick answer is that, having two ears, the brain simply distinguish the provenience of a sound by comparing the intensities and the phase of a sound in the left and the right ear. However, this is surely not enough. To understand why, we simply have to consider the case when a sound has exactly the same intensity, and arrived exactly in the same moment (thus has the same phase) in both ears. Then what? Where is this sound coming from in this case?

As shown in figure 1.15 a sound could be located in completely different positions front, back, top or bottom of the subject and result to have identical input on both ears. We can easily distinguish if a sound comes from those positions, so how does the brain distinguish that? The answer of this is related to the shape of the pinna, the external protrusion of the ear.

まり，脳はまさに時空間解析に基づいてスパイクの連鎖に意味を与えている．この点についてはあとで取り扱う．

同様に，液体中の入れ物には，前述のらせん状の蝸牛殻と呼ばれる器官だけではなく，3つの輪っか状の組織もある．この輪っか状の組織は，同じ外リンパ液で満たされており，直交する3つの軸XYZのように互いに90度の角度で設置されている．これらの輪っかは**三半規管**[50)]という．人間の場合，絶えず頭は動いており，耳管にある液体は常に細毛を揺らすので，振動とその結果としての神経スパイクが起こる．しかし，脳はそれが音ではなく，頭の振動であると判断する．なぜなら，3つの輪っかの管があり，それらが互いに直交しているため，脳はどの方向に頭が動いているのかを判断することができるからである．これは，平衡感覚も生み出す．これは聴覚には関係がないが，偉大なる自然が空間と資源を効果的に使った結果，耳のとても近い部分に位置している．

どのようにして脳が音を精巧に作り上げるのかということを理解するために，人間の**反響定位**[51)]という能力を述べておこう．この機能はとても興味深い．問題は，「どのようにして脳はその音がした方向を認識するか？」である．手短に答えを言うと，耳は2つあるので，右耳と左耳で，音の強さや位相を比較し，音がする場所を認識できるのである．しかしながら，これでは十分ではない．もし左右の耳に届く音が，まったく同じ大きさで，位相もまったく同じだとしたらどうなるだろうか．この場合，音はどこから来ているのだろうか．

図1.15にあるように，音をまったく違う場所である前，後ろ，上，下から発生させ，両方の耳で聞いてみるとどうなるだろうか．私たちは音がどこから聞こえたかを簡単に聞き分けることができるが，脳はどのようにこれを認識するのだろうか．答えは，耳の外側に飛び出た部分，耳介の形にある．どこから音が来たのかを知るために，耳介は複雑で非対称的な形をしていて，違いを聞き分けることができる．特に，聞き取る音の周波数は音が来た方向によって異なる．これは，特定の音質の音は，どこから来たのかによって異なって聞

50) 三半規管：semicircular canals.
51) 反響定位：echolocation.

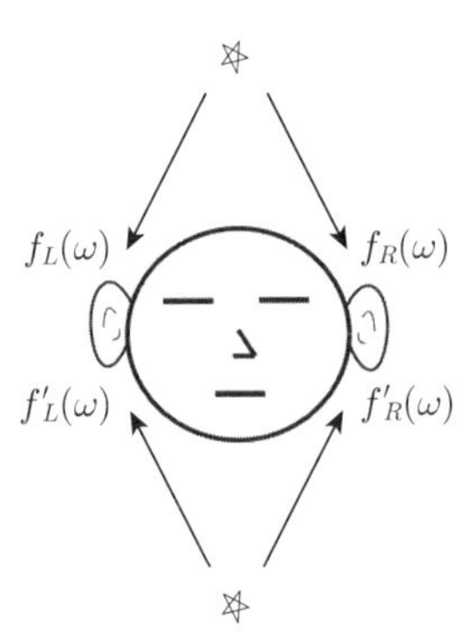

Fig. 1.15 A sound could reach the ear from different directions. Its Fourier spectrum $f_L(\omega)$, $f_R(\omega)$, $f'_L(\omega)$, $f'_R(\omega)$ is changed by the irregular shape of the ear's *pinna*. The spectrum variation depends on the angle of arrival and this helps the brain to localize the source of a sound.

Depending where a sound is coming from, because the pinna has a complex asymmetric shape, the way the sound is picked up differs. In particular, the frequencies that are picked up are different depending on the direction of the sound. This means that the particular *timbre* of the sounds changes depending on where the sound is coming from. The brain is able to discern position of sound source by constant frequency analysis. This quality of the hearing sense depends a lot on the shape of the pinna, but also on the head density and shape as a whole that is generally referred technically with the term *head-related transfer function*, or HRTF. It is possible to measure HRTF of humans simply putting microphones inside the middle ears and measuring the frequency spectrum of test sounds (white noise for example). Reproducing the same frequency characteristics in music recordings, is a method to localize sounds used for example to reproduce the precise position of musical instruments in an orchestra with advanced headphones. Echolocation is very precise in humans, just close your eyes now, and verify by yourself (!)

To understand the very important function of the brain frequency analysis, try this other simple experiment: attempt to change the shape of your

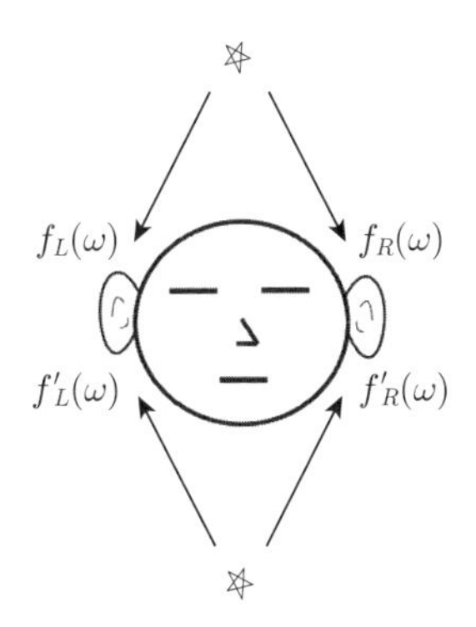

図 1.15 音は違う方向から耳へ到達する．そのフーリエスペクトル ($f_L(\omega)$, $f_R(\omega)$, $f'_L(\omega)$, $f'_R(\omega)$) は耳介の複雑な形によって変化する．スペクトルの変化はどこの方向から音が来たのかに依存し，脳が音がどこから発せられているのかを特定するのに役立つ．

こえるということである．脳は一定の周波数の解析をして，音源の位置を判断することができる．この音の方向を解析する精度は，耳介の形だけではなく，頭の厚みや形などにも依存し，これは，いわゆる**頭部伝達関数** (head-related transfer function: HRTF)[52] と呼ばれる．HRTF は，中耳にマイクロホンを入れて，試験音（例えばホワイトノイズなど）のスペクトルを測定することで簡単に測ることができる．録音したものと同じ周波数特性を再現することは，高機能なヘッドホンを用いて，オーケストラの楽器の位置も再現できる[53]．人間の場合，この反響定位は実に正確である．今，目を閉じて試してみるとよい．

脳の周波数解析という重要な機能を理解するためには，簡単な実験をしてみるとよい．あなたの耳の形を少し変えてみれば良いのである．これはとても簡単で，手のひらを耳の上にかぶせてみるだけでもよい．すると，驚くほどすべての音が歪んで，不明瞭になる．周囲の自然な音がうるさくなり，ホワイトノイズのように変化する．この現象について，これまで学んだことを元に，説明してみてほしい．

52) 頭部伝達関数：head-related transfer function: HRTF.
53) 高機能なヘッドホンは，ある音は後ろから聞こえたり，また違う音は右から聞こえたりするように，左右で周波数を変えて，オーケストラの臨場感を楽しめるようになっている．

pinnas. It is easy, just use the palm of your hands over your ears. You will notice how amazingly all the sounds around you distort and become fuzzy, natural environmental sounds become noisy and morph to sort-of white noise. Now, please try to explain this on the basis of what we just said.

1.9 Vision

How does vision works? To understand this, it is better to understand few things about light. Light is an electromagnetic wave, this means that, like sound, it is a variation of a field in time. In the case of sound, the thing that varies is the air pressure, and the variation is rather slow, from few up to tens of thousands cycles per seconds. In the case of light, the variations are very fast, in the order of 10 followed by 12 zeros cycles per second (amazingly fast!) and the field that variates is electric (and magnetic). Since the light moves at the speed of light $c = 3 * 10^8$ meter per second, at this frequency the wavelength is (see equation 1.5) $\lambda \approx 0.5$ micrometers. A very small size compared to sound waves (!)

The question is: how is this tiny electromagnetic wave detected by our body? Well, we have a specialized organ called *retina* in the back of our eyes. In there we have receptor cells that do exactly that: they detect electromagnetic waves of the right wavelength range. Actually, simply thinking, we have four species of them: the first three types are placed in the center of the eye with high density, they provide sharp and colorful vision. They are called *cones* that recognize long wavelengths (*red*), middle range wavelengths (*green*) and shorter ones (*blue*). In addition to those we also have a fourth type of cells called *rods*, much more abundant in number, but distributed in a larger area, so less dense than the cones. They provide less resolution and they do not recognize colors, but they are more sensitive. If you walk in a full moon night you are probably using a lot of your rods!

Since we have three cone cells sensitive to three colors (red, green and

1.9 視覚

視覚はどのようにはたらくのだろうか．視覚について学ぶ前に，まず光について学ぶべきだろう．光は電磁波である．つまり，音のように場を振動させているということである．音の場合は，空気の圧力を振動させていた．そしてその変化はゆっくりで，1 秒間にせいぜい数万回の振動である．光の場合，その振動はもっとずっと速くなり，1 秒間に 10 兆回もの回数で（驚くほどの速さ！）電気的・磁気的な場を変化させる．光の速度は $c = 3 \times 10^8$ m/s なので，その波長は式 1.5 から，$\lambda \approx 0.5\,\mu$m とわかる．音と比べると，とても小さな値である．

このような極めて速い振動を，私たちの体はどのように感知しているのだろうか．目の裏側には**網膜**[54]という特別な器官がある．この網膜で，受容体細胞は，実際に光の波長の範囲で，電磁波を直接検出している．少し単純化して考えると，私たちは 4 種類の細胞を持っている．3 種類は目の中央に高密度で存在し，くっきりとしたカラフルな視覚をもたらす．これらは**錐体細胞**[55]と呼ばれ，長波長（赤色），中波長（緑色），そして短波長（青色）を認識することができる．4 つ目の細胞は**桿体細胞**[56]と呼ばれるもので，数は多いが，広い範囲に存在するので，錐体細胞より密度は低い．これは解像度が悪く，色も認識しないが，錐体細胞よりも敏感である．満月の月夜の道を歩くとき，あなたはたくさんの桿体細胞を使っているだろう．

錐体細胞は 3 種類の色（赤，緑，青）を認識するので，なぜディスプレイやモニターが 3 つの色 (RGB) をもつのかがわかるだろう．これらの 3 原色の

54) 網膜：retina.
55) 錐体細胞：cones.
56) 桿体細胞：rod.

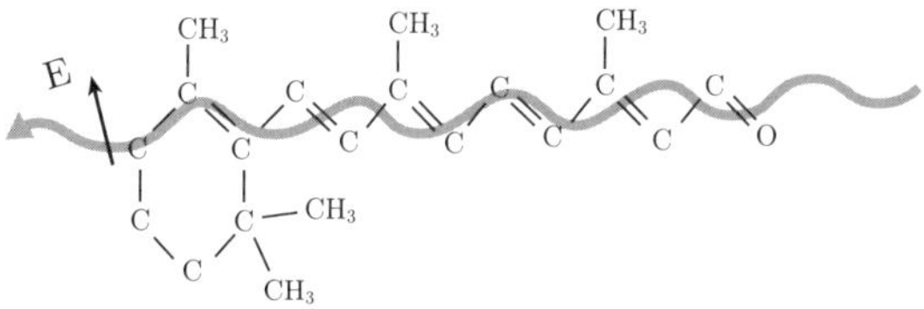

Fig. 1.16 The structure of retinene, this molecule has a long chain of double bonds that are easily stimulated by light waves of the correct wavelengths.

blue), you understand why there is all this fuss about RGBs displays and monitors. Composing the light intensities around these three colors, we can produce any possible color[3].

In these receptor cells the molecule react to light. As said above, the light is an electromagnetic field so it may induce currents on a conductor. In the retina there are no conductors, but the cones and rods contain a molecule called *retinene* that has a particular structure, a long chain of double bonds spaced in a way that makes it sensible to light wavelengths (see figure 1.16). In a double bond, electrons are free to move around, so the chain of the retinene behaves like a sort of conductive material. For this reason, when a wave of the right wavelength arrives, an electric signal reaches the cell body and generates a spike.

Even if the process involved are completely different, the final result is the same: the cell depolarizes and fires a spike, a final action identical to all the other receptors we studied here.

You may ask: "Yes, but how does the eye realize high level pattern recognition?". Of course we do not know the answer to this yet. However, you should realize that simple pattern pre-processing can be realized directly on the retina. In fact, experiments on the horseshoe-crab compound eye have shown that the response of single retina cells is regulated by simple rules

3) People tend to think that composing red and green makes yellow because this is a physical property of the light. This is wrong, red and green makes yellow because of how our retina works! Nothing to do with physics, a lot to do with biology.

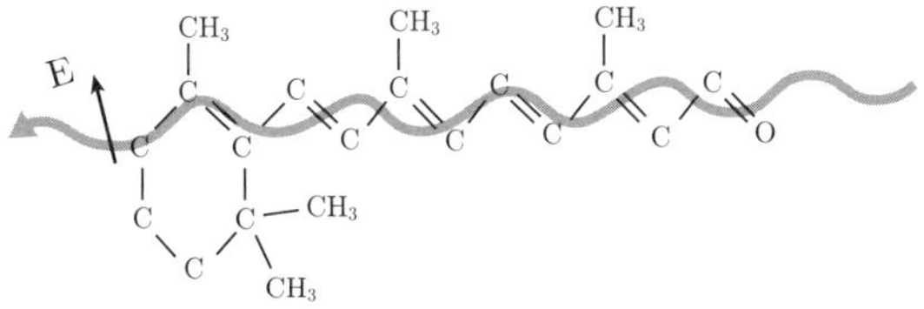

図 1.16　レチナールの構造．この分子は二重結合の長い鎖をもっていて，特定の波長の光によって容易に刺激される．

光量を組み合わせることにより，私たちはどんな色も描くことができる[57]．

これらの受容体細胞内の細胞は光に反応する．前述のとおり，光とは電磁場を振動するので，導体には電流が発生する．網膜は導体ではないが，錐体と桿体はレチナール（レチネン）[58]と呼ばれる物質が含まれている．レチナールには二重結合の長い鎖が両側に存在していて，光の波長に敏感な構造をしている（図 1.16）．この二重結合では，電子は自由に動けるので，レチナールはある種，導体のようにふるまう．そのため，光の波が届いたとき，電気的な信号が受容体細胞に到達し，スパイクを生み出す．途中の過程はまったく異なるのだが，最終的な結果は一緒である．細胞は脱分極し，スパイクが点火する．最終的な形態はどの受容体もみな同じである．

あなたは「光を感じる仕組みはわかった．でもどのようにして目は複雑な形を認識するの？」と質問するだろう．もちろん，まだそれには答えていない．しかし，単純な形の認識の前処理は網膜において直接行われていると気付くことができるだろう．事実，カブトガニの複眼の実験では，1 つの網膜の細胞は単純な法則によって統制が行われ，ふちを認識することが明らかになった．目のなかの受容体細胞は，光の照射だけではなく，周りの受容体の活性にも依存し，異なる方法で活性化する．それらを 4 つの状況にまとめることができる．(a) 対象の細胞は光を照射されていて，周りの細胞は照射されていないとき，対象の細胞は非常に活性化する（明確な活性化）．(b) 対象の細胞が周りの細胞とともに光照射されているときは，細胞は活性化するが，(a) よりは活性化

57) 赤色と緑色を混ぜると黄色になるのが，物理的な性質によるものだと考えがちだが，それは間違いである．赤と緑を混ぜると，網膜の作用により黄色になる．物理の法則ではなく，生物学的な法則によるものである．

58) レチナール：retinene.

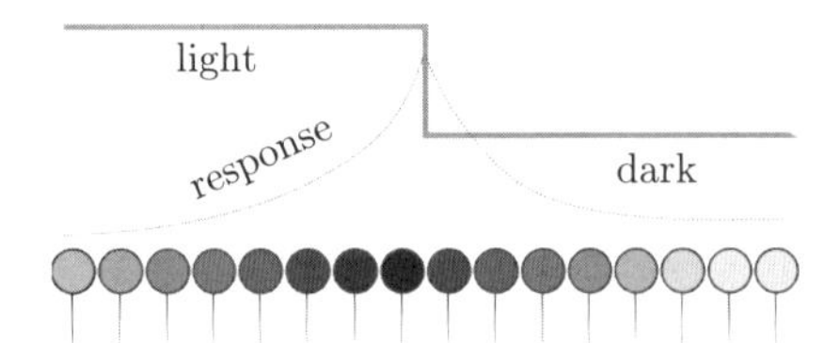

Fig. 1.17 If light illuminates the eye receptors in a step-like fashion like in the figure, their output depends on the light intensity and on the activity of the neighbor receptor. When the illumination level differs between neighbors, the spike rate is higher than when it is similar. This behavior realizes a simple edge detection system. In the figure, receptors in the retina are schematized by the circles and their spike rate by the shade of gray and the dotted line.

that realize edge recognition. The eye receptor cells are activated in different ways depending not only on the illumination, but also on the activity of neighbor receptor. We can summarize four different conditions: a) the cell is under illumination and neighbors are not, the cell becomes very active (specific activity), b) the cell is under illumination with its neighbors, the cell becomes active, but less than the previous case (common activity), c) the cell is not illuminated but its neighbors are, the cell is completely inactive (specific silence) and d) the cell is not illuminated as also its neighbors, the cell becomes inactive, but not completely as in the previous case (common silence).

With these simple four rules, edge detection is automatically realized (see figure 1.17). We can understand how computational algorithms that perform complex tasks as edge recognition, can be actually realized by very simple cellular rules. This is a universal property of our cell-based brain. Superb!

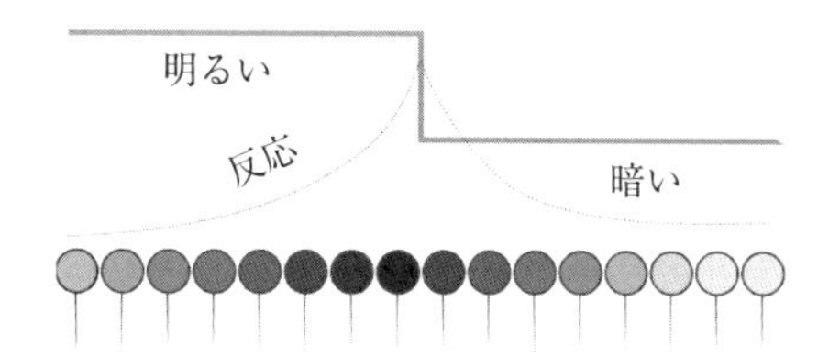

図 1.17　光が図のような階段状の明暗で目の受容体に到達すると，その出力は光の強度と隣の受容体の活性の度合いに依存する．隣と明るさの強度が違う場合，同じ場合と比べてスパイク率は高くなる．この反応は，ふちを認識するシンプルなシステムを実現する．図では網膜の受容体を丸で表しており，そのスパイク率を色の濃さと点線でも示している．

されない（普通に活性）．(c) 対象の細胞に光が照射されていなくて，周りの細胞は光照射されているとき，細胞は完全に非活性となる（完全に非活性）．(d) 対象の細胞も周りの細胞も光が照射されていないときは，細胞は非活性だが，(c) ほどは非活性ではない（普通に非活性）．

この単純な 4 つの規則に従って，物体のふちは自動的に認識される（図 1.17）．ふちの認識などの複雑なタスクをこなすコンピュータアルゴリズムが，実際はシンプルなルールによって実現されていることがわかるだろう．これは脳の細胞が一般的に備えている特性である，

Think about these questions

- Is a cell like a battery?
- What is the function of ATP?
- What is the *rate coding*?
- Why does the spike flow along the axon only in one direction?
- How does the *cochlea* distinguish sound frequencies?
- How does the *equilibrium* sense work?
- What happens if we change the shape of our ear *pinna*?

考えてみよう

- 細胞と電池は似ているか？
- ATP の機能とは何か？
- 発火率とは何か？
- なぜスパイクは軸索の一方向のみ伝搬できるのか？
- 蝸牛殻はどのように周波数を識別できるのか？
- 平衡感覚はどのようにはたらくのか？
- 耳の形を変えてみると，何が起こるか？

Chapter 2

The brain: how does it work

2.1 Introduction

In the previous chapter we studied how the information flows in a living system. The receptor cells are specialized to sense particular signals. These could be temperature or pressure in the skin, or sounds, or visual stimuli or anything else. When the signal is detected by a receptor, the cell becomes *active*, in other words it *fires* spikes of electricity. These spikes flow through nerves up to the brain. In higher animals (like mammals or other vertebrates) the brain deals with an incredible amount of information coming through the nerves from many different areas of the body. These complex signals come simultaneously and are elaborated in real time with an astonishing efficiency. Just think about a simple animal action like, for example, a cat capturing a bird. The cat brain is able to elaborate the complex signals coming from its retina and recognize the moving prey from the background. This simple natural behavior is an extremely difficult task. In fact, if we compare the hardware of modern computer, the retina of a mammal is rather rudimentary. The image of a moving object in the center of focus is captured by only about 100000 photo-receptor cells that are responsible for higher resolution vision (cones from the fovea region of the retina). These data are not constant ones or zeros, but trains of randomly occurring electric *spikes*. These are stochastic in nature, changing in time and shifting in retinal position continuously due to fast movement of the prey and the predator eyes as well. Moreover, data are updated relatively slowly (remember that a neuron has to rest to recharge for at least 5 msec

第 2 章

脳のはたらき

2.1 はじめに

第 1 章では，情報というものが，生物のなかでどのように伝わっているのかを学んだ．情報はそれぞれの特定の信号に特化した受容体細胞で形成される．肌を押す圧力や温度，音や視覚の刺激など，様々な情報である．受容体細胞で信号が検知されると細胞は活性化する．言い換えると，電気的な点火であるスパイクが起こる．このスパイクは神経を伝わって流れ，やがて脳へと到達する．哺乳類や脊椎動物の場合，脳は体の様々な場所から到達した膨大な量の情報を取り扱う．これらの複雑な信号は同時にやってきて，驚くほどに効率よくリアルタイムで処理される．

例として，鳥を捕まえる猫の場合のような単純な動物の動きについて考えてみよう．猫の脳は，網膜から来る複雑な信号を処理することができ，背景から獲物が動いていることを認識する．この簡単で自然なふるまいは，実はすごく難しい仕事である．事実，現代のコンピュータのハードウェアと比べると，哺乳類の網膜のほうがむしろ原始的である．焦点の中心の動いている物体の像は，たった 10 万個の光受容体細胞によって捉えられ，その像は高い解像度（網膜にある網膜中心窩にある錐体）で得られる．これらのデータは，常に 1 または 0 ではなく，ランダムに起こるスパイクが連なったものである．自然のなかの確率論的な信号は，すばやく動く獲物と狩人の目によって時間の経過とともに網膜の位置が連続的に変化する．さらに，それらのデータは，比較的ゆっくりと最新のものになる．神経細胞は 1 回のスパイクにつき 5 ms は再充電に時間がかかることを思い出そう．つまり，脳は 200 Hz の機械ということになる．言い換えると，どのようにして猫の脳が，そんなゆっくりとした原始的な機械をつかって，そんなに上手くかつ簡単に機能するかはよくわからな

after each spike, this makes the brain a 200 Hz machine!). In other words, we have no idea how a cat brain can perform so well and so easily with such slow and apparently rudimentary hardware. Please notice also that the brain does not seem to consume so much power. Neurons spike for few msec, then they rest generally for long, few seconds if they are not active. When neurons are instead active (or very active, which is rare) they still need at least 5 msec of rest to be able to spike again. So, to put is simple, the brain does not consume so much energy because its computational unit, the neuron, is on average inactive. And when it is active, its electric action is very sharp and quick (that's why the action potential is called a *spike*!).

In conclusion, a cat brain consumes just few watts of power, but it can do the amazing job of recognizing, chasing and then capturing a very fast pray. The human brain does similar things too, and does also other much more difficult things that lead to thinking, speaking and generating abstract concepts. We are creative, we have a sense of consciousness, all of these things with 100 billion of neurons running on a total of only about 15 Watts of power, amazing indeed!

We do not know exactly how the brain does these things, but we begin to understand the general principles by which it works. In this chapter we will discuss briefly about these principles.

2.2 Temporal patterns

First of all, we have to admit that the brain is an electric computational machine. Nothing more than that. All that we see, all that we experience and all that we imagine and think is actually happening in our brains. Even reading this book is a computational process happening in your brain. The kind of signals flowing in our brains are very simple: streams (or *trains*) of randomly occurring electric spikes. The spikes signals are nearly identical to each other, so they could be thought as the "ones" and "zeros" of a digital computer. However, one of the most striking differences between

い．脳はまた，それほどパワーを消費しないことにも着目してほしい．神経細胞のスパイクは数ミリ秒かかり，その後，一般的には活性化しない場合には数秒間は休む．神経細胞が活性化した場合（あるいは，稀だがとても活性となる場合），最低でも再びスパイクを起こすのに少なくとも 5 ms はかかる．つまり，簡単にいうと，計算機の単一体である神経細胞は，平均化すると不活性であるために，脳はそれほどエネルギーを消費しないのである．そして，活性化したとき，その電気的なふるまいはとても鋭く速いためでもある（だからスパイクと呼ばれているわけだが）．

猫の脳は数ワットしか消費しない．しかし，その認識は驚くべきはたらきをし，とてもすばやい獲物を捕まえる．人間の脳も同様であるが，もっと難しいこと，例えば考えることや話すこと，そのほか発想を生み出すことなどをすることができる．私たちは創造的であり，うちなる意識をもち，そして一千億個もの神経細胞をたった 15 ワットのエネルギーではたらかせることができる．なんという低いパワーだろうか．

私たちは，どのようにして脳がこれらのことを処理するのかはわからないが，しかし，そのはたらきの一般的な原理は理解した．この章では，それらの原理について詳しく述べる．

2.2　スパイクの時間的な特徴

まずはじめに，私たちは脳を電気計算機，コンピュータであると認めなくてはならない．それ以上のものではない．私たちが見るすべてのもの，私たちが経験できるすべてのもの，そして，私たちが想像したり考えたりするすべてのことは，すべて脳で起こることである．本を読むことですら，脳で起こっているコンピュータプロセスなのである．脳へと送られてくる信号の種類はごく単純なものである．電気的なスパイクのランダムな連続である．スパイク信号は，どれも似通っていて，つまり，1 か 0 かというデジタルの信号と一緒である．しかしながら，脳とコンピュータの著しい違いとしては，スパイクが決定

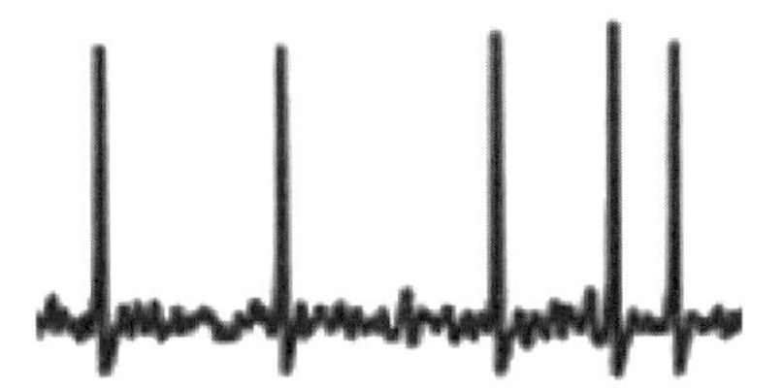

Fig. 2.1 A recording of neurons firing. The spikes are all similar to each other and their timing is random.

brains and computers is that the spike is not a deterministic process. This means that they occur at random intervals[Hong *et al.*, 2012] (see figure 2.1 as an example of real neuronal recording). So, if the signals are random, what is the information? Does the brain work by random calculations? Is it a stochastic machine? Maybe yes, but we cannot confirm that. However, many suggest that the information coded in the brain is the *firing rate*. This means that it is not the single spike timing that transport information, but it is the frequency by which many spikes arrive. In other words, how many spikes are fired per second is the information, not the single spike itself. In this sense, if this firing-rate coding is actually used in the brain, the brain does not need to know exactly when a spike arrives, but only how much is the firing rate of a certain neuron. This is not an established fact at all, it is still an open problem subject to debate among scientists. However, firing-rate is one of the prominent hypothesis about information coding in the brain.

So, how is it possible that vision, smell or touch which are very different from one another are coded in the same way? How does the brain recognize different kinds of signals? Well, the signals are trains of spikes, so they have a temporal character that probably is mainly related to the firing frequency. But, spike trains have also spatial properties: signals from the auditory system are coming from neurons located in the auditory area of the brain, visual signals are spike trains coming from the eyes, etc. The brain deals with very similar signals, but the spatial characteristics (the positional

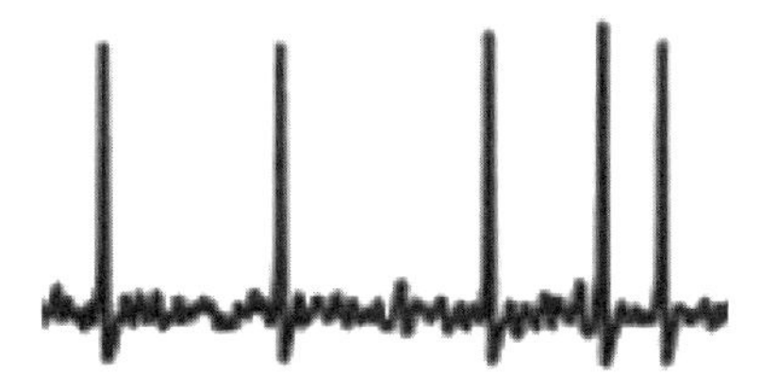

図 2.1　神経細胞の発火率を記録した例．スパイクはどれも似ていて，そのタイミングはランダムである．

論的なプロセスではないということである．これはつまり，図 2.1 に示したようにスパイクがランダムな間隔で起こるということである．

信号がランダムに来るとき，情報とは何だろうか．脳はランダムな計算機としてはたらいているのだろうか？　確率論的な機械として？　おそらく答えは「そのとおり」なのだろうが，それを確かめることはできない．しかしながら，多くの研究者たちが，脳は点火の頻度を情報コードとしていることを提案している．これは単独のスパイクではなく，たくさんのスパイクが到達したときの頻度が情報に変換されることを意味している．つまり，1 秒あたり何回スパイクが点火されたかが情報であり，単独では情報にならないということである．もしも点火の頻度コードが脳で使われているとするこの考え方では，脳はいつスパイクが来たかは重要ではなく，どれだけ来たかを知ればよい．これは実はまだ証明できていない問題で，研究者たちが取り組んでいる問題である．しかしながら，点火の頻度が脳の情報コードであるというのが，今のところ一番有望な仮説である．

それでは，単なるランダムなスパイクの連なりという同じ方法から，視覚，嗅覚や触覚などの感覚が互いに違うことが可能になるのだろうか．脳はその違いをどのように認識するのだろうか．これは，スパイクの連続した信号が，点火の頻度に関係した時間の性質がそれぞれ異なることによるのだろう．しかし，スパイクの連続は，空間の性質ももっている．聴覚のシステムからやってきた神経信号は，脳の聴覚の分野につながっている．また，眼から来る視覚の信号が届くところは，脳の視覚の部分に…といった具合である．脳はとても似ている信号を取り扱うが，信号がどこから来るのかという空間的な性質は，脳にとってとても重要である．それこそが，スパイクの連続が何を脳に伝えてい

location) of these signals is a very important information for the brain. It tells the brain what these trains of spikes mean! In these sense the brain is called by some authors[Hawkins, 2004], a *spatio-temporal* computational machine. To summarize we can say that:

- Vision, Audio, Smell, Taste, Touch... all are *streams* of spikes from neurons.
- Streams are coming from sensory districts in the brain.
- The brain deals with *patterns* of spikes in space and time.

Here a *pattern* is a configuration of signals in time (a series of electric spikes distributed in time) and in space (a certain spatial distribution of active neurons in space). We will discuss this in more details in the following chapters, so for now please focus on this fact: the brain works by elaborating trains of spikes that contain temporal and spatial information. The temporal information could be its spike rate, phase or temporal correlation, in brief time patterns. The spatial information could be the geometrical distribution of active neurons in the brain, in other words: patterns in space. In an even more brief expression, the brain is a spatial-temporal *pattern machine.*

2.3 Brain is a simple and uniform machine

For about a century, scientists have been studying about the brain, and although we have a lot of detailed data, we do not have a large scale theory of it. Actually, we do not have even a basic theoretical framework. We do not know what mathematics tools we need to understand the brain. Many scientists and neuroscientists believed that the brain is too complicated to understand, and so complex that it is impossible to construct a theory of. However, in 1978 the Johns Hopkins University's neuro-scientist Vernon Mountcastle wrote a research paper introducing a new idea: the neocortex, that is the external part of the brain where all the intelligent processes are located, is uniform and simple in nature. Mountcastle's idea goes beyond the pure anatomical observation that the neocortex is uniform. For the first

るのか，ということである．このことについて，Hawkins らは時間・空間上のコンピュータマシンであると述べている [Hawkins, 2004].

まとめると，以下の通りである．

- 視覚・聴覚・嗅覚・味覚・触覚・・はすべて神経細胞からのスパイクの流れによるものである．
- そのセンサーからのスパイクの流れは，脳のそれぞれの区画から届く．
- 脳は，スパイクの時間と空間的な場所の特徴に対処する．

これからその詳細を議論するので，今は脳は，時間や位置の情報を含んだスパイクの精巧な連続によってはたらくことのみに焦点をあてたい．時間の情報というのは，スパイクの頻度であったり，位相やその相関であり，これが明瞭なパターンとなる．空間的な情報というのは，脳の中で活性化する神経細胞の位置的な情報である．もっと明確な言い方をすると，脳は空間的・時間的な特徴を解析する機械なのである．

2.3 単純で均一な機械である脳

ここ 100 年くらい脳を研究している研究者の悩みは，多くの詳細なデータはあるにも関わらず，脳についての一般的な理論が確立されていないことである．事実，私たちはその基礎的な理論の枠組みさえまだ作れずにいる．私たちは脳を理解するためにはどのような数学的な道具をもてばよいのかも知らない．多くの科学者や，神経科学者たちは，脳は理解するのには難しすぎることを知っていて，脳の理論を導くことは不可能であると知っている．しかし，1978 年に，ジョンズ・ホプキンス大学の神経科学者であるバーノン・マウントキャッスルは，新しい考え方を紹介する論文を書いた．それは，すべての論理的な思考がなされる部分である**新皮質**[1]という脳の外側の部分が，本質的に

1) 新皮質：neocortex.

time, Mountcastle suggested the idea that different areas of the brain are doing the same computational operations. If the anatomy is the same, also the computational load and algorithms should be the same everywhere.

This idea seems to be in contradiction with the established fact that different areas of the neocortex are specialized to do different things. The visual system is in the back of our brain, thinking and language in the front, on the side there are auditory areas etc. However, in the previous chapter we learned that the brain communication signals are identical. Any sense such as touch, vision, smell etc. signals its output by streams of electric spikes from specialized receptors. The receptors are different, but the signals they produce are identical. In this sense, it is possible that in every different areas of the brain a *common algorithm* is actually processing the data.

This idea is confirmed by many experiments on animals. For example, experiments on the ferret have shown that removing surgically an eye from its position and connecting it to the auditory nerve, after a period of adaptation, the animal was able to see again (!) The ferret was able to respond to light correctly, demonstrating that the auditory area of its brain was able to somehow learn to process visual signals. This supports the idea that there is a common algorithm that works in all the areas of the brain, independently from its specialization [Roe *et al.*, 1993, Hawkins, 2004]. This astonishing result shows that Mountcastle hypothesis was credible: the brain is uniform and the same common algorithm is processing different kind of data. Other multiple tests on animals and humans show that the brain is *plastic*. This means that any area can adapt to different tasks: the functions of an area that is injured can be moved to other healthy parts of the brain that learn to perform the same functions. To summarize we can say that:

- The neocortex is uniform.
- Different areas of the brain deal with different sensory inputs, but with identical signals.
- There is a common algorithm that works on all the brain.

均質で単純であるという理論である．マウントキャッスルの考えは，新皮質が均一であるという単純な解剖学的な知見を超えている．マウントキャッスルは初めて，脳の違う部分でも同じコンピュータ処理をしているという考え方を提案した．もしも解剖学的に同じであれば，計算量やアルゴリズムもまたどこでも同じになるであろう．

この考え方は，脳の後ろの部分は視覚のシステムがあり，脳の前のほうは言語による思考を司る部分であり，脳の横のところは聴覚に関係する，などといった，新皮質が場所ごとに異なることに特化しているという事実と矛盾しているように思える．しかしながら，前の章で，脳の伝達の信号は，どの感覚（触覚，視覚，嗅覚など），どの信号でも同じものであり，ある感覚に特化した受容体からの電気的なスパイクの流れを取り出すだけだということを述べた．受容体は異なるが，それによって生まれた信号はどれも同じである．つまり，脳のあらゆる異なった部分は，同じアルゴリズムでデータを処理することが可能であることを意味している．

この考え方は多くの動物実験によって確認されている．例えば，外科的に目を取り除き，そこに聴覚の神経をつないだフェレットの実験では，適応期間ののち，驚くべきことに，再び目が見えるようになったのである．このフェレットは光に対して正確に反応し，聴覚を司る部分の脳が，どうやったのかはわからないが，視覚的な情報を正確に処理できるようになったのである．この結果は，脳は共通のアルゴリズムではたらいており，また，それぞれ独立しているというアイディアを支持するものである [Roe *et al.*, 1993, Hawkins, 2004]．この驚くべき結果は，脳は均一で，共通のアルゴリズムで様々な異なる種類のデータを処理しているというマウントキャッスルの仮説に説得力があることを示している．他の複合的な動物や人間への実験は，脳は柔軟な対応をすることがわかっている．これはつまり，どんな領域も違う任務に順応することができ，つまり，傷ついた脳の箇所を取り除き，他の健康な部分に移すと，同部分が同じ機能をこなすことができるということである．

まとめると，以下の通りである．

- 新皮質は均一である．
- 脳の異なる部分は異なるセンサーからの信号を受け取るが，その信号は同

- The brain is *plastic*, any area can learn to perform multiple functions.
- Experiments on animals and humans confirm this.

If the brain is uniform and simple, and if a common algorithm is running in it, then maybe in the future it will be really possible to reproduce it and produce truly artificial brains.

2.4 Brain uses memory to solve problems

Researchers thought that the brain solves problems using complex computational processes. But with the development of advanced theories about numerical computation, it is now clear that the brain is too slow to process the enormous amount of information that it receives. The brain is continuously flooded by information of all kinds. Visual, auditory, touch, smell and all the other information that is constantly streaming in our brain. We now know that the neuron, the single processing unit we have, is a slow device. If so, how is it possible to process such complex information so quickly? And how can it solve such a difficult problems so quickly? Let's consider a very simple problem: to catch a ball. Even a child can do that, but it is a really difficult task. To solve it we need a camera (our eyes). With it we have to 1) recognize the ball from the background. This sub-problem is already very difficult by itself. Once that is solved 2) we have to estimate the distance between the ball and ourselves. We need to use stereoscopy and also get positional clues from the background. 3) We need to have the sense of time to measure the duration of the ball flight in order to estimate the speed of it. 4) Then, we still have to estimate the path of the object and predict its future positions. To do that we need to know gravitation and Newtonian mechanics! Only if we solve 1, 2, 3 and 4 with precision, we can attempt to actually move our hands and body to catch the ball.

To be honest, wouldn't you find it difficult to solve this problem? However,

一のものである．
- 脳はどれも同じアルゴリズムで動いている．
- 脳は柔軟で，どの部分も，様々な機能を遂行するように学ぶことができる．
- 動物や人間への実験がこの考えを支持している．

もしも脳が均一で単純であるとすると，そして，共通のアルゴリズムで動いているならば，将来人工的な脳を再現することは可能になるかもしれない．

2.4　脳は問題を解決するのにメモリを使用する

多くの研究者はかつて，脳は複雑なコンピュータ処理を行って問題を解いていると思ってきた．しかし，理論の発展とコンピュータ計算技術の発展とともに，現在，脳は受け取っている膨大な量の情報を処理するためには処理速度が遅すぎるということがわかってきた．脳はすべての種類の情報が，連続的に流れている．視覚的な情報や，聴覚の情報，触覚，嗅覚といったすべての情報は脳に常に流れ込んできている．私たちは今，神経細胞とは，私たちがもっている1つの単位のプロセスであって，とてもゆっくりとしたデバイスであることを知っている．ではどのようにしてそんな速くて複雑な情報伝達が可能となるのだろう？　そして，どのようにしてそんな難しい問題を瞬時に解けるのだろう？

ここで「ボールを受け取る」というとても簡単な問題を考えてみよう．これは子供にもできることだが，実際に問題として解いてみようとすると，とても難しい．この問題を解くためには，私たちはカメラ（または眼）を必要とする．(1) そのカメラを使い，まず背景からボールを認識する必要がある．この下準備がすでに難しい．そして，それができたとして，(2) 自分からボールまでの距離を見積もらなければならない．立体像を得なければならないし，同時に背景からその位置を測らなければならない．(3) 私たちはそのボールの速度を見積もるために，飛んでくるボールの時間を見るという感覚もなければならない．(4) これらすべてに成功したとして，私たちはまだ，そのボールがどこに飛んでくるかを予測しなければならないのである．そのためには重力という概念も知らないといけないし，ニュートン力学も知らないといけない．この

the paradox is that to catch a ball is something very easy for anyone who has a brain. Even inferior animals can do similar tasks with much smaller brains than ours. A frog can catch a fly in a msec without effort, and her tiny brain has less than a million neurons (humans have more than 15 billion neurons). How it is possible? Where is Newtonian physics stored, where are the equations of motion? And how can we be so fast in computation, a frog captures an insect in msec, but a single neuron cannot fire faster than 200 Hz...!

Well, recent modern theories of the brain suggest that the brain does not process sensory input to compute and solve complex physics problems. It uses a completely different approach. The brain does not solve problems. It does not need that. The brain has already learned the solution of those problems. The solution is already in memory, and the brain only needs to recall it, so it does not require much time.

In other words, the brain processes the sensory input by just searching a similar or identical problem in the memory. If the problem is recognized, the correct solution (or the correct actions) are recalled and activated. This means that the brain is a memory system!

Well you may ask: if the brain just recall memories, where are these memories coming from? Just consider the catch a ball game again. A small child cannot catch a ball, he has to *learn* it! You have to train children in doing anything, such as playing with balls, riding a bicycle, walking, reading or doing more complex things like playing the piano. It may take years to learn to play a Beethoven's sonata, but when it is learned, it is there. It can also take years to learn how to catch a ball, but then we can do it at the end.

The brain memorizes things in a smart way: many experiments confirm that the brain works with sequences of events. What happens if we memorize something, for example a song? Suppose that we have been trained well and now we can play the notes of the song with confidence, one by one. Now I

問題を精度よく解くためには，(1) から (4) すべてが正確にできて初めて，私たちはボールを捕るために体や手を動かすことを試みることができるのである．

正直なところ，あなたはこの問題を解けと言われたら，難しいと思うのではないだろうか．しかしながら，ボールを捕るというのは，脳をもっている人間にとってはとても簡単であるという矛盾がある．人間よりもずっと小さな脳しかもっていない下等動物でさえ，同様の行動はできる．例えば，カエルですら難なくハエをミリ秒の速さで捕まえることができる．カエルの小さな脳には，神経細胞が 100 万個もない（人間の場合，150 億個ある）．そんなほんの少しの脳細胞でどうしてそれが可能なのだろうか？　どこにニュートン力学が記憶されていて，どのように運動方程式を解いているのだろう？　どうして私たちはそんなに早く計算ができるのだろう？　カエルがミリ秒でハエを捕まえるのに，1 つの神経細胞は 200 Hz よりも速くは動けないというのに！

最近の脳の理論では，脳はセンサーからきた情報を処理して，複雑な物理の問題を解いているわけではないとされている．どうやら，まったく違うアプローチをしているようなのである．実は，脳は問題を解いていない．なぜなら，解く必要がないからである．脳は，すでにどのように問題を解くかということを学習しているのである．その解答はすでにメモリの中にあり，脳はそれを呼び出すだけでよいので，それほど時間はかからないのである．

言い換えると，脳は情報が送られてきたら，メモリから似たような問題を探し出しているという処理をしているのに過ぎないのである．すでに問題が認識されている場合には，その解答（または正解となる動き）を呼び出し，活性化しているのである．つまり脳はメモリを使うというシステムなのである！

あなたはきっと，脳がただメモリを呼び出すだけだとしたら，そのメモリはどこからくるのか？と質問するだろう．再び飛んできたボールを捕るという問題について思い出してみよう．小さな子供はボールを捕ることができない．子供はそれを，学習しなくてはならないのである．小さな子供は，何もかも練習し，学習しなくてはならないのだ．ボールを捕ること，自転車に乗ること，歩くこと，もちろん，読むことやもっと複雑なピアノを弾くといったこともそうである．ベートーベンのピアノソナタをピアノで弾くためには何年もかかる．

Fig. 2.2 We can recognize an incomplete pattern because our brain associates the input with known patterns experienced in the past.

ask you: play it backwards! Clearly it is very difficult. Or I can ask you: play the song from the middle. This is also hard. We can play the song easily from the beginning, but it is harder to play it from a random point in the middle of it. Why does this happen? The cause is hidden in the fundamental working principles of the brain: we remember the sequence, but if the sequence is broken, that becomes very hard. Can you say your name backward? If you never did, it will take effort to do that. This means that somehow the brain memorizes transitions from a state to another state. It memorizes sequences of things, and even it memorizes sequences of sequences (like words, sequences of characters, or a poem, sequences of words). This is another outstanding property of the brain functioning.

One of the most important characteristics of the way the brain memorizes things is the so called *auto-associative* property. This means that the brain is able to associate external sensorial stimuli (that is streams of neuronal spikes) with others that are already stored in the brain. This is called the auto-association, because it is the association with something that is already known from the past experience. A typical example of auto-association is how we recognize incomplete images (see figure 2.2). A person that we know well is recognized even if his/her face is partially hidden, or if we see only part of her body. We are able to recognize shapes even if only an incomplete part of them are reproduced. Many behavioral and psychological experiments on humans and animals have proven also that we do not have a huge database of shapes in our brain. Auto-associative memories experiments indicate that the brain uses past experience to recognize what to do and to solve difficult problems.

Another interesting properties of the brain is that we do not memorize all

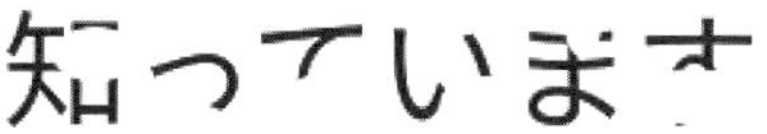

知っていま

図 2.2 私たちは全体が見えていなくても，過去の経験によって知っていれば脳の自動連想機能によって補完される．

しかし，練習すればできるようになる．ボールを捕るのにも何年も練習する必要があるが，できるようになるのである．

脳は賢い方法でこれらを記憶していて，脳は一連の事象について順序通りにはたらくということが，これまでの経験より明らかである．私たちがその記憶を使うとき，何が起こっているのだろうか．例えば，歌う場合には？ 歌について，たくさん練習をして，自信をもってある歌を歌うことができる場合について考えてみよう．このとき，順番を変えて最後から歌えと言ったらどうなるだろうか？ 明らかに，とても難しい．あるいは，途中から歌えと言った場合はどうか？ これも大変である．私たちは，最初から歌うことは簡単ではあるが，途中のランダムな場所からといった場合は難しくなる．どうしてそうなるのだろうか．その理由は，脳の基本的なはたらき方による．私たちは順序を覚えているのであって，その順序が破綻すると，とても難しくなるのである．あなたは自分の名前を逆さまに言うことができるだろうか．一度もやったことがなかったとしたら，それは少し努力しなければできないだろう．これはつまり，脳はある状況から次の状況へどのように移り変わるかを覚えていることを意味している．脳が順序を記憶しているということは，つまり，言葉の場合は文字の順番を覚えていて，詩の場合には単語の順番を覚えているということである．これが実に突出した，脳の機能の特徴である．

脳が順番を記憶しているというやり方で最も大切な点は，いわゆる「自動連想」である．これは，脳は外部のセンサーからの刺激（ニューロンのスパイクの流れ）をすでに脳に蓄えてあるものに関連づけることを意味している．これが自動連想で，過去の経験から知っていることがらを関連づけるためである．典型的な例を図 2.2 に示す．

よく知っている人の場合，その顔の一部，あるいは体の一部が隠されていても，その人だと気がつくことができる．私たちは，その形が不完全であっても，その不完全の部分を再生して認識することができる．人間や動物の多くの

Fig. 2.3 Can you tell these are sheep? Probably yes, because in your brain there is an *invariant* representation of a sheep so you can recognize them even if they are seen from different angles and/or distances.

the possible angles and point of view of objects like our dog, our home, or our car. Moveover, we are able to recognize various objects.

We can recognize a cat by just seeing a cat, any cat. Even a cat that we never saw in our life. This is an incredibly interesting property.

This shows that we can recognize dogs and cats easily, but we do not have a database of all the dogs and cats we have seen in our life, so we are not searching through the database to find the one we are looking at. We can say "this is a cat", even if we have never seen that particular cat (see figure 2.3). This is called the *invariance* property of brain memory. When we look at a cat, a car of any object, somehow the brain is able to classify important *features* of it (a cat has four legs, a fur, a head, two ears... and so on) and store that features associated to the idea of the cat. Once this is stored and properly associated, the brain achieves an *invariant* representation of the cat. This is the awesome property of the way our brain memorizes things.

Invariance can be obtained with another property of brain memorization: *hierarchy*. Many experiments on animals and humans have put forward the fact that the brain memorizes things in a hierarchical way. This means that features of object are somehow put in order of importance. There are important things that are memorized higher in a hierarchical scale and less important things, that are lower. You have to imagine a tree structure, when a pattern is memorized, for example the perception of the image of a cat. Lower in the tree structure many different details are memorized,

図 2.3 あなたはこの写真を見て，これらは羊であると言えるだろうか．おそらく，言えるだろう．なぜなら，あなたの脳には羊の不変的なイメージがあるため，どんな角度や距離で見ても，これらが羊であると認識できるのである．

行動や心理的な経験は，私たちの脳には，莫大な量のデータベースがあるわけではないこともわかっている．自動連想のしくみは，脳が過去の経験を利用して難しい問題を解いていることを示している．

もうひとつ興味深い脳の特徴は，自分の犬や家，車などについて，私たちはすべての角度から見た形を記憶はしていないことである．さらに言うと，私たちは様々なものを見て，それが何かを認識することができる．例えば，私たちは猫を見たとき，どの猫も「猫だな」と認識することができる．たとえその猫を人生で一度も見たことがなかったとしてもである．これはとても興味深い特徴である．私たちは犬や猫を見てたやすく認識するが，もちろん私たちの脳には，すべての人生で見た犬や猫のデータベースがあってそのデータを探しているわけではない．

私たちは，これまで見たことのない個体の猫であっても「これは猫である」ということができる（図 2.3）．これはいわゆる，脳の記憶の普遍的な特性によるものである．私たちは猫を見たとき，あるいはどんな形であれ車を見たとき，脳はともかくも，猫は 4 つの足があり，毛皮があって，頭と 2 つの耳があり…といった具合にその重要な特徴を分類し，猫に関する見解を蓄える．それが一度蓄えられ，その特徴が関連づけられると，脳は猫に関する不変の再現性を得ることができる．これが私たちや動物の脳の記憶という驚くべき機能である．

不変性は脳の記憶のなかで階層という特徴で得られる．多くの動物や人間の経験は，脳が階層構造を用いて記憶しているという事実を裏づける．これは，物体の特徴は重要な順に整理されていることを意味する．記憶には，その階層

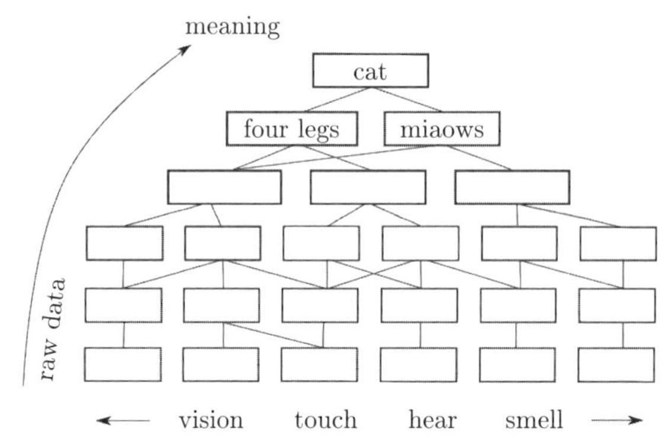

Fig. 2.4 A sketch of a hypothetical hierarchical memory structure. Raw sensory data are at the bottom of the hierarchy. Those are elaborated by brain's functional layers that transmit the results to levels higher up. At each step the data acquire sense. Information flows toward higher brain levels where meaningful stable data are elaborated. Connections and structure of the hierarchy are illustrative.

the single hairs of the fur, some detail of the head, some other particular details. Higher in the hierarchy instead are the facts that this cat has four legs, two eyes, it miaows and so on. Even higher in the brain it is stored the fact that this animal is a cat. Classification of objects done in a hierarchical way (see figure 2.4). The same hierarchical structure of the mind is found in many other cognitive process. We learn music, languages and memorize new words in the same way. Think about how you recognize a song: you listen to few notes and then you say "this is my National anthem!". This does not mean that you recall all the notes one by one. Doesn't even mean that you can sing the anthem or that you can play it. It only means that somewhere in your brain those notes are memorized, because they were heard before. And, sometime in the past, they were associated with the concept of your National anthem (in a hierarchical way the notes were in the bottom of the tree structure, and at the top was the name of the song, in this example the anthem). So we can now summarize some fundamental and amazing properties of the brain:

- The brain uses memory to solve complex problems.
- We remember in sequence (very difficult to play a song backward!).

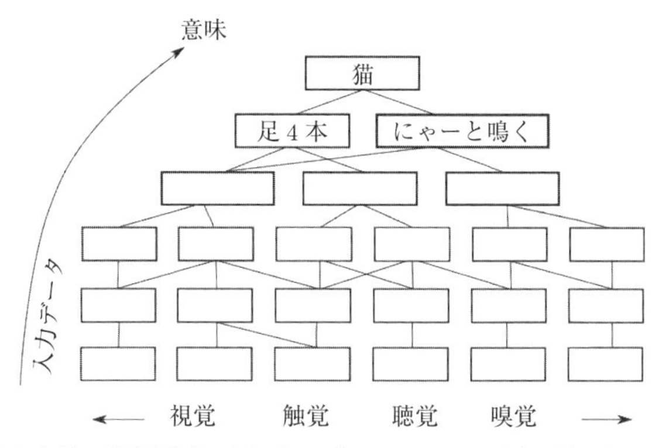

図 2.4　仮想的な記憶の階層構造の図．加工前のデータは階層の低い部分にある．重要性がある情報の場合，それらは高い階層へとレベルアップする．図には猫の場合の例が提示されている．

によって，大切なものと，それほど大切でないものがあり，前者は高階層に，後者は低階層に記憶される．例えば，猫のイメージにおける認識について，記憶のこの階層はツリー構造をイメージしてみるとよいだろう．下位階層では，例えば毛皮の毛がどうなっているとか，頭の形だとか，その他の些細な特徴など多数のディテールが記憶されている．一方で上位階層では，猫は4つ足であること，目が2つあること，にゃーと鳴くこと，などといったことが記憶されており，さらに高い階層では「猫」という概念が記憶されている．図 2.4 のように，物体の分類は階層に分けて行われているのである．

思考の階層構造と同様のものは，他の認識の過程においても多くある．私たちは音楽や言語を学ぶとき，新しい単語を同じ方法で記憶する．歌をどのように認識するかを考えてみよう．例えばあなたはいくつかの旋律を聴いただけで，「これは国歌だな」と気がつく．これは，あなたが1つ1つの旋律を思い出すことを意味していない．また，国歌を歌ったり演奏できるという意味でもない．これはただ，国歌を以前に聞いたことがあるため，脳のどこかにその旋律が記憶されていることを意味する．そして，過去に，それが国歌と関連づけられているのである．階層としては，下位階層ではその旋律が，そして上位階層では，その歌の名前（この例では国歌）が記憶されている．さて，脳の驚くべき，基礎的な特性について，まとめてみよう．

- The brain has a way to memorize things in a hierarchical structure. Details and variable unimportant features at the bottom of the tree. Important features for classification and grouping at the top.
- Hierarchy causes *invariance.* Objects are always recognized as themselves even if details are different (a cat is always a cat: from any angle, of any color!).

2.5 Intelligence and prediction

Some decades ago the first computers able to play chess became available. And now, so called *intelligent* machines are feasible. What do we mean with intelligent? Is a computer that beats me in chess intelligent? This is a very debatable question, the answer depends on the definition of intelligence. Many people agree on the fact that intelligence is somehow defined by behavior. If a machine is able to behave in an intelligent way, that is like a human, then we can say it is intelligent. For example, if a machine can beat a human in chess, we can say it has a certain level of intelligence. In 1950 A.M. Turing published a very interesting paper with the title "Computing machinery and intelligence"[Castelfranchi, 2013], I suggest you to have a look at it. There he supposed that a human judge sits behind a computer terminal asking general questions to a computer and a human located in another room. The test works this way: the judge types on the terminal any kind of questions (like "how are you?", or "what is the square root of 46732?"), and reads the answer on the terminal. Turing claimed that if the judge takes too long to understand who is the computer and who is the human, then the computer is intelligent. This test became very famous and it is now called "the Turing test". Still now, there are no machines that can succeed in the Turing test. After few questions the human judge is able to identify the computer, with high possibility. Nevertheless, this test implicitly shows that the definition of intelligence depends on the behavior

- 脳は複雑な問題を解くために記憶を使う．
- 私たちはその順序を覚えている（歌を逆さから歌うのはとても難しい）．
- 脳は階層構造をつくって記憶する．詳細や変化する特徴は下位階層に，重要な特徴は上位階層に分類される．
- 階層構造は脳が不変性をもつ原因となる．不変性とは，物体は詳細が異なっても，それ自身として認識される（猫はどの角度から見ても，どんな色をしていても，猫だと認識される）ということである．

2.5 知性と予知

数十年前，初めてコンピュータでチェスの対局が可能となった．そして現在，いわゆる知能機械は実現可能になりつつある．しかし，知能とは何を意味するのだろうか．チェスの対局で私たちを打ち負かすコンピュータは知的なのだろうか．これは論争の余地のある問題である．答えは，知能の定義による．多くの人は，知能とはそのふるまいによって定義されるということに同意するだろう．もしも機械が人間のように知的にふるまうとしたら，機械も知能があると言えるだろう．例えば，機械がチェスで人間を打ち負かすことができれば，機械が一定の知的レベルにあると言うことができる．1950 年に A.M. チューリングは「計算する機械と知性」[Castelfranchi, 2013] という興味深い論文を発表した．この論文はぜひ読んでみて欲しい．この論文のなかでチューリングは，違う部屋に設置したコンピュータ端末と人間に質問を行うことで，機械に知能があるかを判断できると提案している．実験は次のように行われた．ご機嫌いかが？　や 46732 の二乗根はいくつか？　などといった様々な種類の質問を片方が行い，ディスプレイ上への返信を読む．チューリングは，コンピュータからの返信か人間からの返信かを質問者が判断するのに時間がかかりすぎる場合には，コンピュータも知能があると判断できるとした．この実験はとても有名で，チューリング・テストと呼ばれる．現在も，チューリング・テストに高い確率で合格できる機械はない．いくつか質問して，すぐにコンピュータであると判断できる場合がほとんどである．つまりこのテストは，知能とはふるまいであることを暗に示している．チューリングテストは，「機械が私のようにふるまえば，それは私のように知的である」ということを述べてい

of the machine. It is like the Turing test says: "if the machine behaves like me, then it is intelligent like me".

This idea was accepted for decades. However modern neuroscientists like Jeff Hawkins[Hawkins, 2004] have shifted the attention from behavior to *prediction*. According to Hawkins, intelligence is defined not by a particular behavior, but by the ability to predict future events. The idea is that the neo-cortex's main activity is to predict the future evolution of certain stream of data. Since the 1990, words like *inference*, *prediction* and similar expressions became popular, why? This is because the effort to make intelligent machines forced scientist to study prediction and inference. The way we understand the world is actually based on prediction. There are some examples: if you open a door, you expect to hear known sounds when the door moves and when it closes back behind you. When you touch the door, you expect to feel something by your hand, and you expect the knob to offer a certain resistance and to be in certain position. You may ask "well, actually when I open a door, I do it automatically, I do not even realize that I am opening the door, I am not predicting anything at all". It is a good question, but we know that our brain is continuously predicting things because if something not expected happens, then we immediately notice! For example, suppose you are going to open the door of your room like you do so every day, but somebody changed the knob. Now the knob is in another place, it is not exactly where it should be, your brain will immediately signal it to you: "wow what happen to the knob, somebody moved it?". This is true for any sensorial feeling. If the door does not make the usual noises you will say "hey, somebody oiled my door?". Anything we do is constantly associated with a predicted sensorial reaction. Our brain sees the world as a constant stream of input data associated with a stream of expected reactions. When something is new we pay attention, but all the other things are dealt automatically by our brain, the brain expects the result of any motor or perceptual event, but if something differs from the expected behavior it

る．

この考えは何十年もの間受け入れられてきたが，現代の神経学者，例えばジェフ・ホーキンス [Hawkins, 2004] は，ふるまいから「予想・予測」へとその注目すべき点を変えてきている．ホーキンスによれば，知能とは，特定のふるまいによって定義されることではなく，将来起こることを予測できるかどうかであるとしている．このホーキンスの考え方とは，端的に言うと「新皮質が膨大なデータの流れから将来の展開を予測するのにはたらく」ことである．1990 年代から推論や予測など，似ている意味をもつ単語が一般的になったが，それはなぜだろう．それは知能機械をつくろうと努力して，科学者たちは予測と推論を学んできたためである．私たちが世界を理解する道は実際予測に基づいている．ここに，いくつかの例がある．あなたがドアを開けるとき，ドアを開けたときにする音や，後ろに閉めるときも既知の音を確認するだろう．ドアに触れたとき，あなたは手から，何かを触るという感覚を予期するだろう．そして，ドアノブからの一定の抵抗を感じることを予期するし，ある一定の位置にあることを予期するだろう．あなたはおそらく「でも，実際にドアを開けるとき，私はそれを自動的にやっているし，ドアを開けているという実感はないし，何も予想などしていない」と言うだろう．それはいい疑問である．しかし，実際のところ，私たちの脳は，絶えず予測をしている．なぜなら，もしも予期しないことが起こった場合，それに直ちに気がつくためである．例えば，あなたは毎日のように部屋のドアを開けるだろう．しかし，誰かがノブを変えたとしよう．ドアノブは今，違う場所にあって，どこにあるかわからない．あなたの脳は直ちに「ドアノブに何かが起きたんだな，誰かが動かしたのだろうか？」と信号を送るだろう．これはどんな感覚にもあてはまる．もしもドアがいつも通りの音ではなかった場合，あなたは「おや？　だれかドアに油をさしたのかな？」と思うだろう．私たちの行動は，常に予測した感覚反応と結びついている．私たちの脳は，常に連続的なデータを読み込んで予期している反応があると思って世界を見ている[2]．何か注意すべき新しいことがあり，その他

2) うるさい部屋で寝ているときはうるさい音が連続的にインプットされているので，急に静かになると起きるという経験はないだろうか．これと同じで，脳はこれまでインプットされていた情報と同様の情報が入ってくると予想しているということである．

is immediately noticed by higher levels of the hierarchy and our attention is focused on that.

Where are all those predictions coming from? Are they somehow calculated? No, modern theories of the brain suggest that the brain does not calculate predictions according to models of the world, the brain just creates a huge catalog of past events, memorized in a hierarchical structure. So when we open a new door, our brain stores all the sensorial experience hierarchically and then this sensorial experience is recalled and played back next time that we open the same door. If everything goes well, all expectations are satisfied and we do not notice anything special when we open that door again and again. The information about the sensorial experience of opening that door is not important and we do not pay attention to it, however our brain is instead constantly predicting everything down to the smallest detail. So, to unlock the mystery of brain functioning and intelligence, we have to understand the way the brain uses memory and succeed in prediction.

2.6 The structure of neo-cortex

We said previously that the neo-cortex is the part of the brain that surrounds the core of it. The inner part of our brain looks smoother and it is very old. It is responsible for all the automatisms of our body and our animal behavior (fear, anger, etc.). This inner part of the brain is sometimes called the *paleo-brain* or the *lizard-brain*. The neo-cortex evolved recently, that's the reason for the *neo* prefix, the *cortex* suffix stays to indicate that is something that surrounds the paleobrain externally to it. As said before, the cortex is a large sheet of brain tissue, about 50 cm by 50 cm, so large that must be folded to fit in our skull. That's way the human brain looks so curvy and curly.

The cortex is actually made by 6 layers, one on the top of each other (figure 2.5). A surgeon can verify that, just peeling away one sheet over the

はいつも通りの場合には，脳はいつも通りの出来事を予期しているので，何か予期した振る舞いと違うことが起こった場合にはただちにそれに気がつき，脳の階層の高いところで，その違いに対して集中するのである．

その予期とはどこから来ているのだろうか．どのように計算されているのだろうか．現代の脳の理論では，脳は世界のモデルについて予想を計算するのではなく，過去の出来事に関する膨大なカタログをつくっていて，それを階層システムに記憶しているということが支持されている．つまり，新しいドアを開けたとき，私たちの脳はすべての知覚的な経験を階層的に記憶し，この知覚的な経験は次回，同じドアを開けたときに呼び出すことができるようになっている．もしも万事滞りなく進んだとき，すべての予測はあたり，次回，そしてその次もそのドアを開けたとき，特別なことには何も気がつかない．このドアを開けるという知覚的な経験についての情報は重要ではなく，普段は注意をはらっていないが，脳は常に些細なことも予想をしているのである．つまり，脳の機能と知性の謎を解くためには，私たちは，どのように脳が記憶を使っていて，予想に成功しているかということを理解する必要がある．

2.6 新皮質の構造

前述したように，新皮質は脳の中心部の周りにあるものである．脳の中心部は滑らかで，とても古くから動物にある[3)]．それは，例えば心臓が動いたり，肺で呼吸するなどといった私たちの体の基本的な動きや恐怖や怒りといった原始的で動物的な振る舞いなどの反応を引き起こす．この脳の内側の部分は「古い脳」や「トカゲの脳」と呼ばれる．新皮質は最近進化したものであるため，「新」皮質と呼ばれる．皮質というのは，古脳の外側を覆っているために，そう呼ばれた．皮質は脳の組織を包む大きなシートのようなもので，50 cm 四方くらいと大きく，頭蓋骨に収まるように折りたたまれている．そのため人間の脳は曲がり，ねじれている．

3) 古い脳はトカゲの脳ともいう．新皮質は哺乳類などの賢い動物だけがもっている．その中にあるトカゲの脳は原始的な動物にもある．人間もすごく怒ったとき，トカゲの脳が知性を超えてはたらく．これが野生的・いわゆる動物的な感情である．つまり，周りの新皮質は知性を司っていると言えるのである．

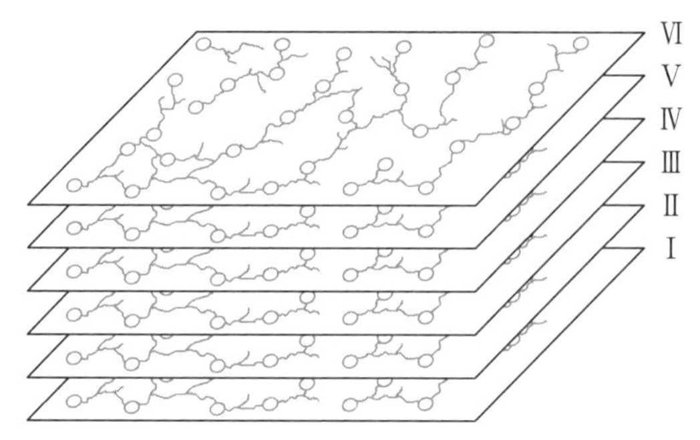

Fig. 2.5 A sketch of the neo-cortex layered structure. Information in layer V or VI are hierarchically higher than the others. They represent more meaningful and abstract concepts than those in lower layers (as you can see in this figure, layers are generally indicated by Roman numerals).

other. This means that there are 6 layers of horizontally connected neurons, and each layer communicate to the next one vertically with a lesser number of connections. If the number of connections vertically and horizontally would be similar, we would not have layers, but a three-dimensional bulk of neurons.

Is this structure of 6 layers important for our understanding of how the brain works? Yes, it is! It has been suggested that the higher levels (the more external in the neo-cortex) are those where general invariant ideas and concepts are stored. Whereas the inner layers are those where variable and detailed information are memorized. There are exceptions, for example the area of the primary visual cortex (called V1) is mostly involved in detecting very simple features (like edges and binocular disparity) in tiny regions of the visual field.

Nevertheless, many experiments on live animals and humans have shown that this hypothesis is supported by the fact that on the lower levels the electrical activity have faster frequency characteristics than the higher levels. This is maybe related on how the brain evolved[Kaas, 2011].

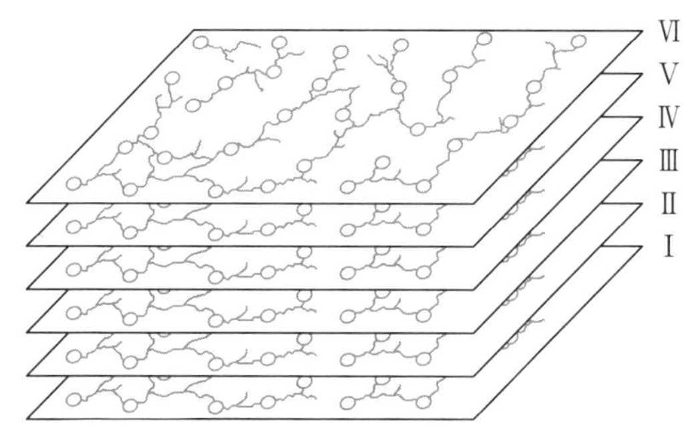

図 2.5　新皮質の構造の模式図．階層をローマ数字で示す．V や VI の層は他よりも高い階層である．これらの層では，階層の下の層にくらべて，より意味があり，概念的な記憶が記憶されている．

図 2.5 に示すように，新皮質は 6 つの層になっていて，重なっている．外科医は，新皮質の 6 層は，それぞれを別々に剝がすことができると証言する．これが意味することは，6 層は水平方向にニューロンが繋がっており，それぞれの層は垂直方向に少しのニューロンでつながって情報を伝達している．もしも垂直方向の伝達と水平方向の伝達が似ているとすると，新皮質は層ではなくて 3 次元的な神経細胞の構造をもつが，そうではない．

この 6 層の構造は，脳がどのようにはたらくのかを理解するうえで重要な意味をもつのだろうか？　実は，これこそが重要なのである．なぜかというと，高い階層（より新皮質にとって外側にある層）は，一般的や不変的な思想や概念を記憶しておくところだからである．それに対して，内側の層は変化する事柄や詳細な情報が記憶されている．もちろん例外もある．例えば，視覚皮質（visual cortex V1）は，高い階層にあるが，物体の端や両眼視差などとても単純な情報を取り扱っている．

しかしながら，たくさんの動物実験から，下の層は電気的な周波数が高く，外側の階層では周波数が低くなることがわかっており，層によって取り扱う情報が違うという仮説を支持している．これは，どのように脳が進化してきたかということと関係している [Kaas, 2011].

原始的な生物は抽象の能力が限られているが，現代の人間や高等な哺乳類は，物事を概念化することが可能で，これは新皮質の進化とともに可能になっ

Primitive animals[1] have limited abstraction abilities, whereas modern humans and higher mammals are able to conceptualize many things. This happens with the evolution of layers in the neo-cortex. Also, in nocturnal primates specialized in visual predation, it appears that the 6 layers of the cortex are subdivided in at least other 12 sub-layers in the visual area of the neo-cortex of these animals. This suggests that indeed layering is important to brain functions. Where specialization and intelligent behavior is needed, more layering is observed anatomically.

Another important feature of the architecture of the brain are *columns* of neurons. Neurons appears to have connections that span across all the six layers vertically. These structures are called columns. In some sense, information flows horizontally through large specialized area of the brain (for example for vision, or for earing or anything else) and then important information is communicated higher in the level. The hierarchical structure seems to be *stratified*, that is information of the same importance and details circulate in the same layer, and information that should be of higher level contents, should go up to next layer, to the next hierarchical level.

As shown in figure 2.4 we could have area of the brain specialized in touch, other in vision, other in earing, for example. These large area of the neo-cortex are represented as boxes in the figure. In these areas, the 6 layers communicate information mainly at the same level of hierarchy, however information is joined and stored at higher levels aggregating the information in a meaningful way. For example, we can learn that when we pour water in a glass we hear also the characteristic sound of the water. This is stored as a higher level concept. In fact, if we pour water in a glass and hear the sound of something else instead of water we will be very surprised. The visual information and the associated sound are aggregated and bounded strongly in our brains.

---

1) Simple organisms with small brains, like insects, worms or small fishes.

たことである．夜行性の霊長目は，暗い中で捕食するために視覚的に特殊化されており，6 層の新皮質が少なくとも視覚の部分については 12 層の複層に分離していることがわかっている．このことは，層状になることが脳の機能に本当に重要であることを示している．特別な機能だとか知性な振る舞いが必要な場合，より層が増えることは解剖学的に観察されている．

その他の脳の構造上の重要な特徴は，神経細胞のカラムである．神経細胞は 6 層の間を垂直に繋いでいる構造があるように見える．この構造をカラムという．いくつかの感覚では，情報は水平方向に伝わり，脳のその感覚に特化した部分に伝わる．例えば，視覚や聴覚などである．そのとき，重要な情報は高い階層へと伝わる．階層構造は層状化されており，これは同じ層の情報の重要性は同様で，詳細は同じ層を循環している．そして，もっと高い階層にあるべきと気が付いた重要な情報は高い階層へと押し上げられる[4)]．

図 2.4 に示すように，私たちは，例えば触覚，視覚，聴覚などそれぞれの感覚に応じて特化した脳の部分をもっている．これらの大きな新皮質の部分は，図の箱のように表すことができる．これらの場所では，6 層の情報の伝達は同じレベルの階層で行われる．しかし，情報は結合し，高いレベルに集められ蓄積される．例えば，私たちはグラスに水を注ぐときに，その水に特徴的な音を学ぶ．これは，高い階層のレベルにて蓄積されるべき概念である．事実，もしも私たちがグラスに水を注ぐときに音を聞き，その音が水とは異なる音がしたら（例えばビー玉だとか金属球が落ちるような音がした場合），私たちはとても驚く．水を注ぐときはこのような音がするだろうという視覚の情報と音という聴覚の情報，2 つの情報は私たちによって収集され，脳のなかで強く結合されているからである．

4) 同じ層には同じ重要性の情報が入っている．音楽を聴いていて，「あ！　あの曲だ！」となったら，その情報が上の層に伝わっている．

2.7 How does neo-cortex learn?

Learning is one of the most important properties of the brain. Since the brain is just a bunch of organized neurons, learning becomes a process of connecting neurons. Neurons are physically connected, anatomically. This cannot change, learning cannot be done by rewiring the brain! Learning happens, because one neuron connects with nearly 10 thousand of other neurons by its branching *axons*, and receives about the same number of input connections from other neurons. The connections between a branching neuron's axon and the input *dendrite* of another neuron is called *synapses*. Each neuron can have more than 10 thousand synapses. The fact relevant to learning is that the synapses connection strength is not constant. It is variable and can get stronger accordingly to particular rules. The most simple rule was discovered not by a biologist or a surgeon, but by the Canadian psychologist Donald O. Hebb in the 1950s. The so called *Hebbian learning* rule is very simple: synaptic strength is increased if two neurons are firing together (correlation) and decreased if they do not (anti-correlation). In other words, if two neurons fire at the same time, the synapses between them gets strengthen, if not it gets weaker. This can also be remembered with the phrase: *fire together, wire together*[Hawkins, 2004]. We now know that Hebbian learning is not enough to explain perfectly what is happening in the neo-cortex. We know that learning is more complex than that, many experiments have shown that our brain runs many variations of the Hebb rule. This means that some synapses can maintain their strength for long time, other can change quickly, other can react to spiking timing in a very sensitive fashion while other can be rather hard to change strength.

Even if the details of the learning processes in the brain are still mysterious, the Hebbian principles can explain most of the cortical functions.

2.7 新皮質はどのように学ぶのか？

脳の機能のなかで最も重要なことは，「学習する」ことである．脳は神経細胞の束にすぎず，学習することを通じて神経細胞をつなぐのである．神経細胞は解剖学的に，物理的に結合する．この結合は変化せず，学習することで再結合することはできないのである．学習すると，1つの神経細胞は近くにある1万個もの他の神経細胞と，軸索の束となって結合し，繋がった神経細胞からの入力の連結を，同じ数だけつくる．この神経細胞の軸索の間の連結と樹状突起はシナプスと呼ばれる．それぞれの神経細胞は1万個を超えるシナプスをもっている．学習に関係する事実は，シナプスが繋がっている強さは一定ではないということである．その結びつきの強さは変化し，一定の規則に従って強く結びつく．最も単純なルールは，生物学者でも外科医でもなく，カナダ人の心理学者であるドナルド・O・ヘッブによって1950年代に発見された．これはいわゆる「ヘッブの法則（またはヘブの法則）」と呼ばれるもので，ごく単純である．シナプスの結びつきは，互いのニューロンが発火すると強さを増し，それがあまり起こらないと強度が弱くなるというものである[5)]．もう一度繰り返すと，2つのニューロンが同時に発火すると，その2つのシナプスの間の結びつきの強度は強くなるということである．そして同時に発火しないと結びつきは弱くなるのである．このことは「fire together, wire together（ともに発火すれば互いにつながる）」というフレーズで記憶されている [Hawkins, 2004]．現在ではこのヘッブの法則は新皮質に起こっていることを完全に説明するのには十分ではないことがわかっている．学習というのは，もっと複雑なものであり，多くの実験は，私たちの脳が様々なバリエーションのヘッブの法則にのっとってはたらいていることを示している．つまり，いくつかのシナプスは長時間強度を保つのに対して，すぐに結びつきが弱くなるものもある．敏感に発火が起こっても，強度が変わらないような結びつきもある[6)]．

5) 試験の前に一夜漬けしたらそのときは覚えているけど，その試験の後にはすぐに忘れてしまう理由はこれである．

6) 何回読んでも忘れてしまう分野の科目もあるし，1回聴いただけで忘れない内容の授業もある．

Think about these questions

- Is the brain faster compared computers nowadays?
- What is a *temporal pattern*?
- What is Vernon Mountcastle idea?
- What does it mean that the brain is *plastic*?
- How does the brain solve complex problems?
- What is the *auto-associative* property of the brain?
- Can you relate the concept of *intelligence* and *prediction*?
- What are the 6 layers of the neo-cortex?
- What is the *Hebbian learning* principle?

脳のなかで，どのように学習という過程が進むのかはまだ謎なところもあるが，ヘッブの原理は新皮質の機能の重要な部分を説明している．

考えてみよう

- 脳は最近のコンピュータと比べて速いだろうか？
- スパイクの時間の特徴について説明せよ．
- マウントキャッスルの仮説について説明せよ．
- 脳が柔軟な対応をすることについて説明せよ．
- 脳はどのように複雑な問題を解くのか？
- 脳が自動的に連想するときの特徴について説明せよ．
- 知性と予測というのはどのように関係づけられるだろうか？
- 新皮質の 6 層とは何か？
- ヘッブの原理とは何か？

Chapter 3

Mathematical models of the neuron

Neurons have been a very well-known kind of cell for many decades. Mathematical models of its functioning have been established, and one of the first successful model was developed by Hodgkin and Huxley. The basics of it (usually named as the HH model) was introduced in chapter 1. It describes the neuron as a non-linear device composed of three different batteries. Each battery represents the so called *reversal potential* which is a difference in various ion concentration within the neuron cell.

In the model, potassium (K) and sodium (Na) are the main currents reproduced. Because these two currents are in competition and they are connected to non-linear impedance, they produce a strong non-linear behavior.

For decades, neural models derived from and developed based on the HH theory and as a result, they have become a very good approximation to the experimental behavior of the single neuron. However, the model is complex and computationally heavy. Is it possible to grasp the real essence of the neuron behavior and make a simple phenomenological model that works?

We have to notice that the neuron is a dynamical system. This means that mathematically it is not linear and can have complex behaviors depending on the input conditions. You have to imagine this experiment: a neuron is kept in vivo, at the right temperature and the proper media. An electrode is placed in its membrane to give electric stimuli and to measure its reaction to them. At the beginning of the experiment, our neuron is at *rest potential* of about $-70\,\mathrm{mV}$ as said in chapter 1 (see figure 3.1). Let's suppose we give a brief stimulus of few mA, then what happens to the neuron? We

第3章

ニューロンの数学的モデル

ニューロンはこの数十年間，細胞の一種として良く知られてきた．ニューロンの数学的モデルは，ホジキンとハクスレーによって確立された．彼らの数学的モデルはニューロンのはたらきを表す好例として高く評価された．その基礎となるホジキンとハクスレーのH-Hモデル[1)]については，第1章ですでに紹介した．そのモデルでは，ニューロンは3つの異なる電位によって構成され，その出力は非線形的にふるまう．3つの電位は，逆転電位[2)]と呼ばれるニューロンの細胞内におけるイオンの濃度差を示す．

H-Hモデルにおいて，カリウムとナトリウムは互いに電流としての役割をもつ．非線形のインピーダンス[3)]に接続されている2つの電流はカップリングを引き起こす．そして，それらは強い非線形性を示す．

数十年の間，ニューロンモデルはH-Hモデルをベースに発展してきた．結果として，それらのモデルは，単一ニューロンの実験結果ととても良く近似している．しかし，モデルは非常に複雑で数値解析的にも負担が大きいという課題が残っている．ニューロンのふるまいの本質を捉えることはできないのだろうか？　また，それらをごく単純な現象として扱うことは不可能なのだろうか？

まず，ニューロンが動的なシステムであることを理解しなければならない．その数学的な意味は，線形ではなく，入力に依存して複雑なふるまいを得ることができるということである．ここで生体内[4)]のニューロンに関する実験について考えてみよう．適切な温度かつ場所にあるニューロンの生体膜[5)]に電極を

1) ホジキン (Hodgkin)-ハクスレー (Huxley) モデル：H-H model.
2) 逆転電位：reversal potential，イオン濃度差による電位.
3) インピーダンス：impedance，電圧と電流の比.
4) 生体内の：in vivo，生化学や分子生物学等で用いられる用語.
5) 生体膜：membrane.

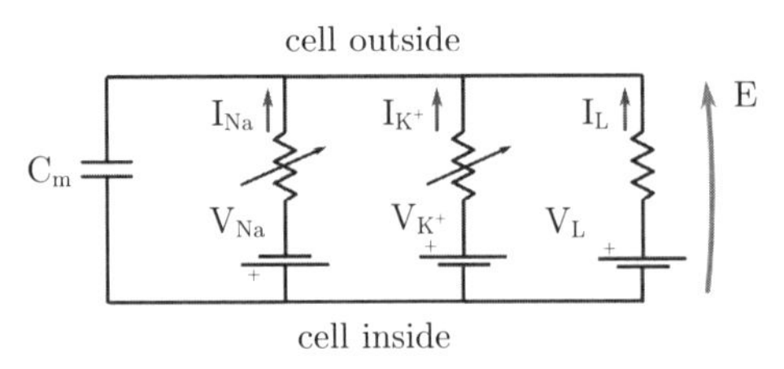

Fig. 3.1 A scheme of the Hodgkin and Huxley model developed in 1952 studying the giant squid axon.

record a tiny increase of membrane potential, that decays soon to the rest potential. However, if we give two of these stimuli very close to each other, we observe a spike of membrane potential. The spike could reach values up to 30 mV, out of a stimulus of only few mA. If we make our stimulus longer, we observe periodic spiking!

How can we explain this behavior in mathematical terms? One way to do that is to use phase diagrams. Let's imagine a diagram where on the horizontal axis we have the neuron membrane potential and the vertical axis the number of potassium K^+ gates open "n".

When the neuron is at rest, the membrane potential is around -70 mV. This happens because of the ion pumps that create a balance between positive and negative ions flowing in and out through the cell membrane. So, plotting in the *phase space* "n" versus "membrane potential" like in figure 3.2, a single point will represent this equilibrium. This point is said to be *stable*, because if we give a small perturbation, like said above a brief impulse of few mV, the cell moves away from that point, but goes back soon to rest at -70 mV. In phase space this is represented by the curve in figure 3.2.

If, we could control the injected current instead, we would notice that for smaller values of current, the neuron will spike and return to rest. However, if the impulse is more intense, after the first spike, the system will remain in an excitable area, and the spikes will repeat periodically. In the phase

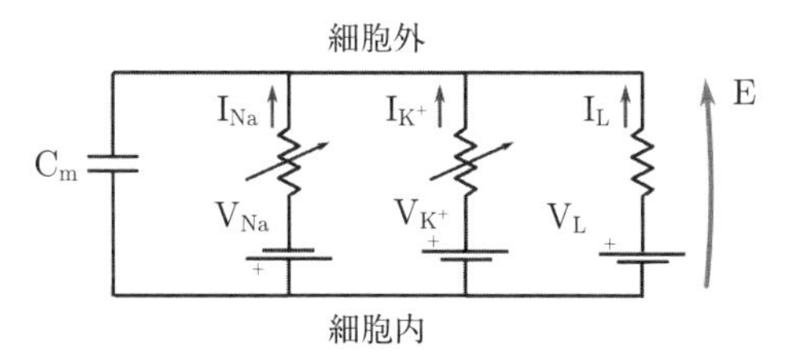

図3.1　上図は，1952年のホジキンとハクスレーによってイカの巨大軸索に関する研究で報告された．

設置して電気的な刺激を与え，その反応を測定した．実験を始めるとすぐに，ニューロンは約 −70 mV の静止電位[6)]と言われる反応を示した．静止電位についてはすでに第1章に述べた通りである（図3.1）．たった数 mA の刺激を与えると，ニューロンには何が起こるだろうか？　膜電位のわずかな増加を記録し，その後，すぐに静止電位へと減衰するだろう．また，もし互いに時間的に近い2つの刺激を与えたとき，膜電位に活動電位[7)]を観測することができる．その活動電位の電位差は 30 mV にまで達する．つまり，与えた数 mA の刺激を大きく上回っていることに気がつくだろう．さらに刺激を与える時間を伸ばすと，周期的な活動電位を観測することができる．

これらのふるまいを数学的にどのように表すことができるだろうか．図3.2に示した状態図を使って説明する．まず，横軸にニューロンの膜電位，そして縦軸にカリウム (K^+) の開かれたゲート数 n を置いて状態図を描く．

ニューロンが静止電位になるときの膜電位は −70 mV 程度になる．これは，細胞膜を行き来する正イオンと負イオンのバランスで生み出されるイオンポンプによるものであり，図3.2に示すようなゲートの数と膜電位の位相空間[8)]を描くことができる．また，その中のある一点は平衡状態を示す．この点は安定状態と呼ばれる．もし，数 mV のパルスからなる摂動を与えれば，細胞はいったんその状態から遠ざかるが，すぐに静止電位である −70 mV に戻ってし

6) 静止電位：rest potential.
7) 活動電位，スパイク：spike.
8) 位相空間：phase space，一般に位相空間とは，1つの質点の運動状態を，位置と運動量を用いて示したものである．ここでは，ニューロンの状態をイオンチャネルの数と膜電位で示す．

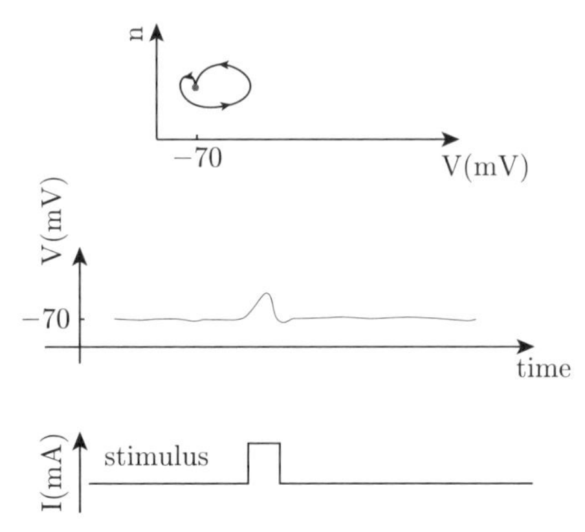

Fig. 3.2 The top graph represents the phase space graph where time runs along the curve in the direction of the arrows. The middle graph represents a membrane potential fluctuation as time passes. The bottom graph represents a stimulation by a brief current. As you can see, a brief current stimulus induces a membrane potential fluctuation. In this example, you can see the cell variables returning to the equilibrium central point, which is represented by the small dot, due to the lack of open channels.

space we will observe an orbit around the stable point, a so called *attractor*, see figure 3.3.

This different behavior (periodic cycles against a single spike) depends on the initial conditions and input current. From the dynamic systems theory point of view, the change of behavior is called a *bifurcation*.

A good neuronal model, should be able to reproduce not only the electrophysiological behavior, but also the bifurcation dynamics of the neuron.

In other words, the neuron is behaving like a non-linear machine. Its output is chaotic and stochastic, but its general behavior is reproducible with dynamical models that we know how to construct.

A well-known and very efficient type of simple model is the *quadratic integrate and fire* neuron model, developed by Izhikevich[Izhikevich, 2000]. Starting from the Hodgkin-Huxley model, Izhikevich reduced the number of variables and derived (see ref [Izhikevich, 2010], chapter 5.2.4) a simplified model that reproduces almost perfectly the neuron behavior and its dynam-

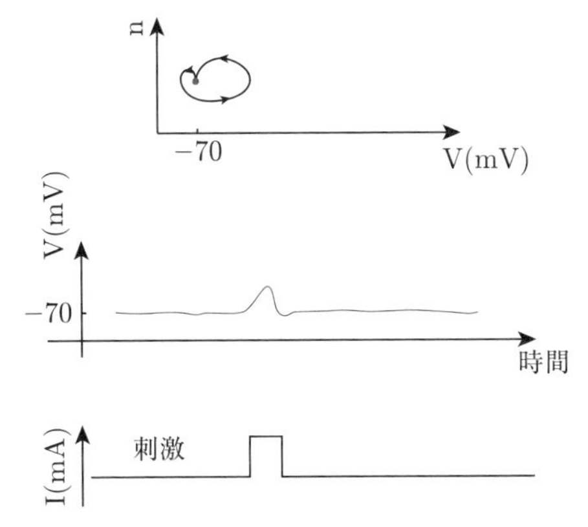

図 3.2　図の上段は，位相空間を示している．図中の矢印は，時間によってどちらの方向に進むかを表している．図の中段は，膜電位の時間変化を示しており，図の下段は単一の電流による刺激を表している．単一の電流による刺激 (図下段) は，膜電位に変化を引き起こす (図中段)．この例では，刺激された時間が短いため，開いたチャネルの数は十分でなく，細胞は平衡状態の中心点に戻る．

まう．それらは図 3.2 に示す，位相空間の曲線で表される．

代わりに，入力電流を制御することができれば，電流の値が小さくなるのを観測することで，ニューロンが静止状態に戻ったことを検知できる．しかし，もし刺激が強すぎた場合，最初の発火現象の後に，そのシステムは不安定な状態を保ち，周期的にスパイクを繰り返す．座標系では，アトラクター[9]と呼ばれる，安定点の回りに軌道を描くことが観察できる．それを図 3.3 に示す．

単一の発火現象と周期的な発火現象の違いは，初期条件と入力電流に依存する．動力学的な観点では，それらを分岐現象[10]と呼んでいる．

ニューロンの的確なモデルとは，ニューロンの電気的な振る舞いだけでなく，分岐現象についても説明する必要性がある．

言い換えれば，ニューロンとは非線形を引き起こす装置である．そして，その信号は無秩序で確率的である．しかし，ニューロンの一般的なふるまいは，構築方法を把握している動力学的なモデルによって再現可能である．

9) アトラクター：attractor，非線形力学における系の時間発展．
10) 分岐現象：bifurcation，パラメーターにより系の挙動が大きく変化する現象．

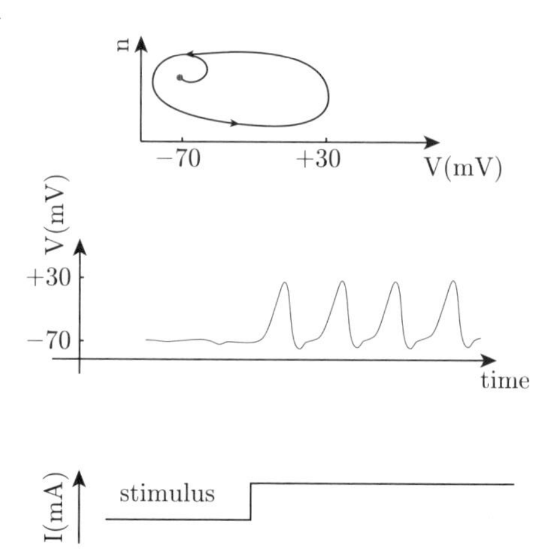

Fig. 3.3 If the stimulus is prolonged, the number of channels opened reaches a threshold and this induces a chain of events that provokes the periodic spikes as in the middle graph. The orbit in the phase space (the top graph above) represents this.

ics. Model results and real electrophysiological data on single neuron, are indistinguishable even by an expert eye.

This model can be represented by two main variables as below, where v is the intracellular membrane voltage and u is the cell recovery potential. The variable I represents the external current. The variable u represents the total effect of the various internal ion currents in the cell.

$$\frac{dv}{dt} = k_1 v^2 + k_2 v + k_3 - u + I \tag{3.1}$$
$$\frac{du}{dt} = a(bv - u)$$
$$\text{if } v \geq 30\,\text{mV then } v \leftarrow c, u \leftarrow u + d$$

The three parameters k_1, k_2 and k_3 are obtained by fitting the dynamics of real cortical neurons, and they are set to 0.04, 5 and 140 respectively. In the expression above, the unit of v is mV and the unit of t is msec[Izhikevich, 2003]. By manipulating the other four parameters a, b, c and d, it is possible to reproduce the dynamical behavior of common types

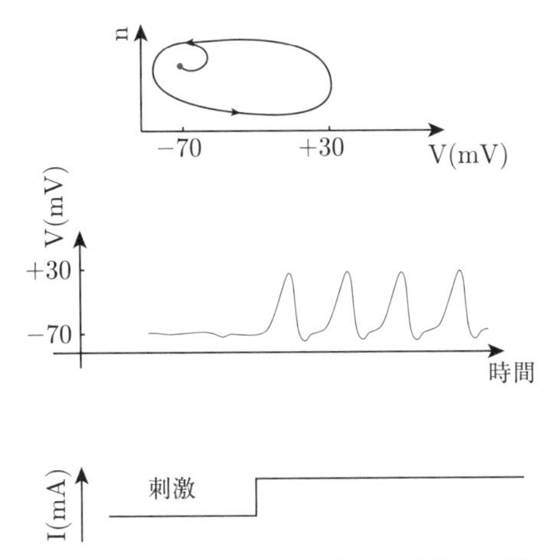

図 3.3　刺激の時間が長い場合の例．上段は位相空間，中段は，膜電位の時間変化，下段は刺激の電流値を示している．長時間の刺激は，開いたチャネル数のしきい値を越え，周期的なスパイクの連鎖を誘発する (図中段)．そのため，平衡点である出発点に戻らず，位相空間に軌道を示す．

簡単で有効なモデルとしてイジケヴィッチ[11)]による積分発火ニューロンモデル[12)]が良く知られている [Izhikevich, 2000]．イジケヴィッチは，H-H モデルから出発し，変数を減らし，式を単純化することで，ニューロンの動的なふるまいを再現することに成功した ([Izhikevich, 2010]，5 章 2.4 を参照)．モデルの結果は単一のニューロンの実際の電気信号と良く一致した．実際それは生物学の専門家から見ても疑いようがなかった．

このモデルは次の 2 つの変数によって表すことができる．v は細胞膜内電位，u は細胞復帰ポテンシャルを示している．また I は電流である．細胞復帰ポテンシャルとは，理論的なもので，細胞内の様々なイオン電流による総合的な影響を考えたものである．

11) Eugene M. Izhikevich：ロシア出身 数学者．
12) 積分発火ニューロンモデル：quadratic integrate and fire neuron model.

of neurons like the so called *chattering* neurons, *fast spiking* neurons, *regular spiking* neurons and *intrinsic bursting* neurons.

Other similar neuronal models exist. They are not strictly representing biologically meaningful variables, but they reproduce their dynamics efficiently, which is their characteristics.

Dynamical models, which are also called *spiking* models, should not be confused with the simpler static models called *artificial neural networks* (ANN) where the neuron is a non-linear device that has no dynamic as explained in chapter 4.

Even if ANN was able to realize very useful systems with deep learning features, interest in spiking neural networks would be very high. You have to consider that a spiking neural network is energetically incredibly efficient. As to the conventional ANN, the output value of each single neuron is at a constant electric level. It must be saved to the computer memory. Many neurons in "on" or "off" state represent a long sequence of ones and zeros that requires small level of currents for each state. Even if these currents are only few mA, to simulate a real brain, we would need millions or billions of those, which is energy-consuming. Also each single calculation must be very fast. Nowadays, computers need to have Giga Hertz of speed to be useful. This means they require enormous currents and energy to run. This is the reason why supercomputers cost so much and consume a lot of energy, because they are based on static systems. A brain simulator based on static ANN will have the same problems.

On the other hand, a biological brain is a spiking neural network. The neuron is always in a sort of *off state* when it almost does not consume energy. It has a brief spike sometimes, with very low frequency and a great number of high hierarchical connections. With this working structure, it may contain a colossal number of neurons that consume a very low amount of energy and doesn't need to be fast. You do not believe this? Our own brain contains about 100 billion neurons (10^{11}!), which run at about 200

$$\frac{dv}{dt} = k_1 v^2 + k_2 v + k_3 - u + I \tag{3.1}$$
$$\frac{du}{dt} = a(bv - u)$$
$$v \geq 30\,\mathrm{mV}\ \text{ならば}\ v \leftarrow c, u \leftarrow u + d$$

k_1, k_2, k_3 は実際のニューロンの動作原理を適合させることで得た値で，それぞれ 0.04, 5, 140 である．また，v の単位は mV，t の単位は ms を使用する [Izhikevich, 2003]．a, b, c, d の 4 つのパラメーターを変更することで，ニューロンの様々なふるまいを再現することができる．それらはそれぞれ，チャタリングニューロン，ファストスパイクニューロン，レギュラースパイクニューロン，イントリンシック[13)]バースティングニューロンと呼ばれる．

イジケヴィッチのモデルの他にも類似のニューロンモデルは複数存在している．イジケヴィッチが提唱したものではないモデルの特徴は，厳密には生物学的な意味がある変数を用いていないところにあるがニューロンの挙動を効果的に再現することが可能である．

動力学的モデルを単なる確率モデルと勘違いすべきではない．この章で説明したようなモデルはニューロンの動力学的な挙動を考えており，スパイキングモデルとも呼ばれる．次の第 4 章で非線形信号発生装置として取り上げる人工ニューラルネットワーク (artificial neural networks：ANN) は時間的な挙動を考慮していない．

ANN が深層学習機能をもった非常に有用なシステムを実現できたとしても，ニューラルネットワークのスパイクに対する関心は非常に高い．私たちは，スパイキングニューラルネットワークのすばらしさについて考える必要がある．従来の ANN では単一のニューロンの電気信号を仮定して，コンピュータに 1 と 0 の繰り返しを記録させる必要がある．私たちの脳を再現するためには，1 つ 1 つに数 mA の電流が必要になる．そして私たちの脳では 1 千万，1 億のとても速い演算が必要になってくる．今日のコンピュータはギガヘルツのスピードで情報を処理している．つまり，脳を動かすために莫大な電流とエ

13) intrinsic：固有の．

Hz and consume few Watts (about 20 in an adult). Well, the brain is a computational machine and we know how a single spiking neuron works, so if we could understand how the brain does calculations, we could build new kinds of computers. Extremely powerful as performance, but very efficient as energy consumption. Imagine: a computer with billions of neurons that consumes so little that can be run by batteries...!

Think about these questions

- What is a *dynamical system*?
- Can you describe the *spiking* of a neuron?
- What is a phase diagram?
- What kind of input can cause the neuron to spike periodically?
- Can you describe the Izhikevich model?
- What is an Artificial Neuronal Network (ANN)?
- What is the main advantage of Spiking neuronal networks as against ANNs?

ネルギーを要することを意味する．これは統計的な視点から見ても，スーパーコンピュータのコストが高すぎることを意味する．つまり，統計モデルに基づくANNにも同様に計算量過多の課題が残る．

逆に，実際の生物学的な脳は，スパイキングニューラルネットワークである．ニューロンには，エネルギーを消費しないオフ状態が存在し，低周波数の発火現象もある．さらに，その数は膨大で，高次元の階層構造をもっている．この構造のはたらきは，莫大なニューロンが少ないエネルギーと遅いスピードで動かしている．これが信じられるだろうか？　私たちの脳内には1000億個のニューロンが存在し，約200 Hzで動き，数ワット（成人の場合で約20 W）を消費している．つまり，脳がコンピュータであり，単一の発火ニューロンのはたらきであることを私たちは知っている．そのためもし，脳内でどのように演算が行われているかが理解できれば，きっと新しい種類のコンピュータを作ることができるだろう．それは，とても強力で高効率であるだろう．数億のニューロンが小さな電池でコンピュータを動かしている未来をイメージしてみよう．

考えてみよう

- 動力学的なシステムとはどのようなものか？
- ニューロンの発火現象を表現してみよう．
- 位相空間とはどのようなものか？
- ニューロンの周期的な発火を引き起こす入力はどのようなものか？
- イジケヴィッチのモデルを表現してみよう．
- 人工ニューラルネットワークとはどのようなものか？
- ANNに対して，スパイキングニューラルネットワークの最大の強みはどのような点か？

Chapter 4

Artificial neural networks

Inspired from the successful theoretical studies on neurons (chapter 3), mathematicians and computer scientists developed new computational tools called *Artificial neural networks* or ANN. However, in these networks the neurons are represented as simple signal integrator. It outputs a number, not a dynamical signal (the "spike"), but a constant value determined accordingly to simple integration rules. The signal coming in from the input is summed up and squeezed to a normalized value between zero and one. The generic equation for an ANN "neuron" is:

$$Y_o(\bar{X}, \theta) = \mathcal{L}(\Sigma_i X_i W_i + \theta) \tag{4.1}$$

Here Y_o is the output signal. $\mathcal{L}$ is the *squeezing* function (usually called *activation function*). This is generally a *sigmoid* (that is any function that has the shape of an "S", see fig 4.1), $\bar{X}$ represents the input received by the neuron (a vector $\bar{X} = \{X_1, X_2, X_3 ... X_i\}$). Similarly, W_is represents the *connection strength* weights between inputs, and θ represents an offset, often called *bias*.

Maybe you think that squeezing is not realistic or important. However, *squeezing* of input values is really occurring in biological neurons. In fact, spikes from previous neurons are summed up, but the resulting action potential is not getting any bigger if the number of connections is bigger. If the number of simultaneous spikes in input to a neuron is high, the resulting action potential does not change in amplitude or shape. All action potentials are similar to each other, despite a changing input. The neuron have a mechanism of normalization that is usually described mathematically in

第4章

人工ニューラルネットワーク

第3章で紹介したニューロンに関する理論研究に感化された数学者やコンピュータサイエンティスト達は，人工ニューラルネットワーク (ANN) と呼ばれる新しい数値計算ツールを発展させた．ANN における個々のニューロンは，ある入力値に対して単一の出力信号の発信源として表される．その出力信号は，発火現象のような動的なものではなく，単なる数値の積算と閾値を用いたルールによって，その値を決定する．

入力値の合計が閾値に到達すると，0 から 1 の区間で規格化された値を出力する[1]．ANN では，一般に単一ニューロンを次式で記述する．

$$Y_o(\bar{X}, \theta) = \mathcal{L}(\Sigma_i X_i W_i + \theta) \tag{4.1}$$

ここで Y_o は出力信号を示す．$\mathcal{L}$ は活性化関数[2]であり，一般的には活性化関数にシグモイド関数[3]を用意する．式 4.1 はシグモイドの頭文字 S に関する等式を示す．$\bar{X}$ はニューロンへの入力を表している（入力 $\bar{X}$ は，ベクトル表記で $\bar{X} = X_1, X_2, X_3, ...X_i$ を意味する）．また，ベクトル W_i は入力間の結合強度として重み付け[4]を示す．そして，θ はバイアス[5]と呼ばれ，入力にオフセットを与える．

もしかしたら皆さんは，活性化関数の定義に，現実味や重要性を感じないかもしれない．しかし，活性化関数は生物学的知見からもニューロンのシナプスのふるまいに一致し，事実として発火現象はニューロンへの入力の積算によっ

1) 規格化：normalize，出力の要素数は 1 になるので，入力の要素数を圧縮する操作が必要となり，これを規格化という．

2) 圧縮関数：squeezing function，直訳は圧縮関数だが，ANN 分野では活性化関数を意味する．本書では活性化関数と表記する．

3) シグモイド関数：sigmoid，神経細胞のモデル化に用いられる関数．

4) 結合荷重：connection strength weights，重み付けを結合荷重と呼ぶ．

5) バイアス：bias，バイアスまたは偏り．

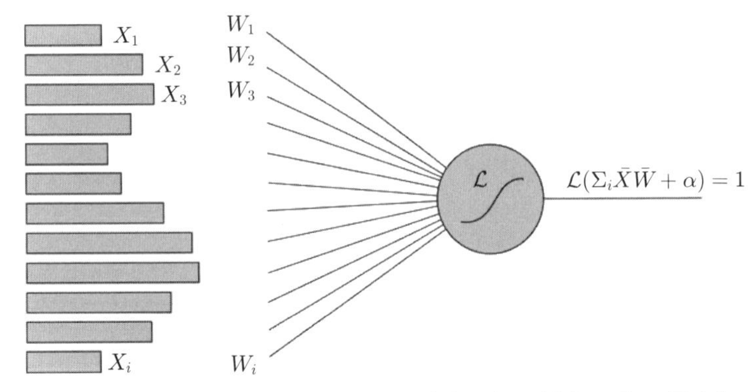

Fig. 4.1 The sketch of an artificial neuron and the sigmoid function $\mathcal{L}(x)$ (a *perceptron*). The X_i values are the inputs and the W_i values are the connection weights. The length of the bars visually represents the input numerical values. When trained, the perceptron is able to recognize and classify the input patterns.

ANN by a logistic function. Considering $\bar{X}$ and $\bar{W}$ as vectors, equation (4.1) can be rewritten in matrix format using the *dot* product:

$$Y_o(\bar{X}, \theta) = \mathcal{L}(\bar{X} \cdot \bar{W} + \theta) \tag{4.2}$$

4.1 The perceptron

What can you do with the *artificial neuron* in figure 4.1? A single neuron like this can be trained to learn things! Suppose to have many inputs $\bar{X} = X_1, X_2, X_3, ...X_i$, the same number of connections weights $\bar{W} = W_1, W_2, W_3, ...W_i$ and a desired neuron output Y.

We give the neuron many examples of $\bar{X}$, let's call those examples $\bar{X}_n$ and the corresponding desired output Y_n. As a practical example of this: $\bar{X}_n$ could be blood tests values and the Y_n the corresponding positive or negative results of a medical exam. The $\bar{X}_n$ are the *inputs* and Y_n are the

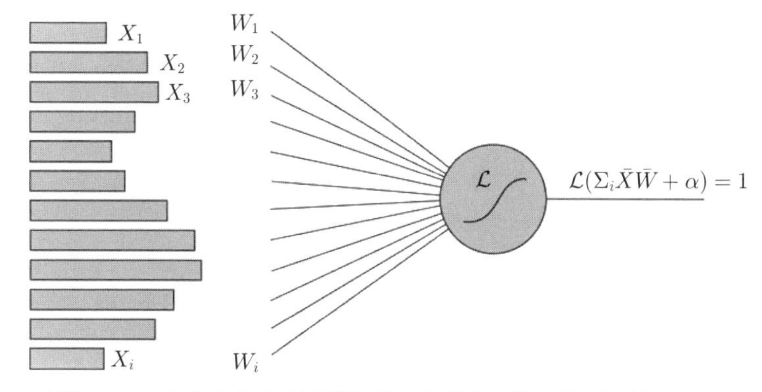

図 4.1 人工ニューロンとシグモイド関数 $\mathcal{L}(x)$ を表す．単一パーセプトロン X_i は入力，W_i は結合荷重を表す．棒グラフは入力値のパターンを表す．学習を終えたパーセプトロンは入力値を認識，分類することができる．

て引き起こされる．つまり，もし入力が複数存在しても活動電位に達しなければ発火現象には至らない．多くのニューロンが同時に発火する場合にも，活動電位は変わらない．活動電位は単なる入力というわけではなく，増幅器に近い役割を果たしている．ANN において，ニューロンの数学的記述は，ロジスティック関数による正規化のメカニズムとして用いられる．式 4.1 において，$\bar{X}$ ベクトルと $\bar{W}$ ベクトルは内積を用いて行列で書き直すことができる．

$$Y_o(\bar{X}, \theta) = \mathcal{L}(\bar{X} \cdot \bar{W} + \theta) \tag{4.2}$$

4.1 パーセプトロン

図 4.1 の人工ニューロンを用いて何を扱えるだろうか？ 単一のニューロンは学習を行うだろう．今，$\bar{X} = X_1, X_2, X_3, \ldots X_i$ は入力，$\bar{W} = W_1, W_2, W_3, \ldots W_i$ はそれと同数の結合荷重[6]，Y は出力と仮定する．

次に，ニューロンに複数の教師データ $\bar{X}$ を与える．それらを $\bar{X}_n$，得られる出力を Y_n とする．より具体的な例を挙げると，$\bar{X}_n$ は血液検査の値で，Y_n は検査結果の陽性／陰性を示す．$\bar{X}_n$ は入力，Y_n は検査項目に対応する．

6) 結合荷重：connection weights.

corresponding *labels*.

Believe it or not, we can train this neuron to *learn* these examples and even to *predict* the results of new, unknown tests by just optimizing the connection weights W_i! A neuron of this type is sometimes called a *perceptron*, why? Because this single mathematical entity is able to learn the numerical patterns at the input, and reproduce the desired results. Not only that, properly trained (that is with properly adjusted $\bar{W}$) perceptron is even able to output a prediction, given an input. Smart!

Ok, so how can we find the optimized $\bar{W}$? It is easy: we define a so called *cost function*, that is a function that represents how good the perceptron result is. This function can be, for example, the *squared error* SE:

$$SE = \frac{1}{2}(Y - Y_o)^2 \tag{4.3}$$

For every example n that we present to the perceptron, we have to change all the elements of $\bar{W}$ in order to reduce the cost function. We can do that this way:

$$W_i = W_i - \alpha \frac{dSE}{dW_i} \tag{4.4}$$

Now, let's apply the equation above to calculate each element i of $\bar{W}$. In simple words: the above equation 4.4 means that we modify each element W_i of $\bar{W}$ by an amount proportional to how much that element influences the cost SE. The derivative dSE/dW_i tells us how much we have to change. This method to optimize $\bar{W}$ is named *gradient descent* because the parameters $\bar{W}$ goes down (descend) a slope (the gradient dSE/dW) to reach the minimum error SE. Sometimes the cost function SE is also called the *energy* of the system.

The parameter α represents the average *descent speed.* This value is important and so should be chosen carefully, not too big otherwise we may skip over our minimum, not too small otherwise we end up in a local one.

Of course the problem here is to calculate the exact gradient formula. In

実は，複数の教師データを使ってニューロンを学習させることで，思いもよらぬ結果を導くことができる．それは，結合荷重 W_i を最適化することで得られる．このようなニューロンをパーセプトロンと呼ぶ．その理由は，入力値パターンの学習と結果の出力を数式化できるところにある．さらにパーセプトロンは $\bar{W}$ を学習によって最適化することで，与えられた入力から予測を出力することも可能である．

次に実際にどのように $\bar{W}$ を最適化するのかを見てみよう．まず，目的関数[7)]と呼ばれる関数を定義しよう．これは，パーセプトロンによる出力がどのくらい適していたのかを示す関数と言える．ここでは，具体例として誤差の2乗 (squared error:SE) の関数を扱ってみよう．

$$SE = \frac{1}{2}(Y - Y_o)^2 \tag{4.3}$$

次に目的関数の値が減少するように結合荷重 $\bar{W}$ を変化させながら，パーセプトロンに教師データ n を与えなくてはならない．結合荷重の最適化の方法を次に示す．

$$W_i = W_i - \alpha \frac{dSE}{dW_i} \tag{4.4}$$

上記式を結合荷重 $\bar{W}$ の要素 i 個について計算する．簡単に説明すると，式4.4は $\bar{W}$ の各要素 W_i を誤差関数を用いて適正値に修正することを意味する．つまり，微分 dSE/dW_i によって結合荷重の値を修正する．この方法は，最急降下法と呼ばれ，$\bar{W}$ を勾配 dSE/dW_i で最小誤差 SE になるように修正していく．その意味で，目的関数 SE を系のエネルギーと呼ぶことがある．

変数 α は，平均勾配速度を表す．その値はとても重要なので慎重に選択する必要がある．なぜなら，もし値が大きすぎると最小値をスキップし収束せず，もし値が小さすぎると局所的な解に収まってしまうからである．

もちろん何か実際の問題を考える場合には，微分方程式を正確に解く必要がある．実際，出力 Y が結合荷重 W に依存しない単純な問題に対しては微分 dSE/dW_i を計算することができる．

7) 目的関数：cost function.

our simple case we have to calculate the derivative dSE/dW_i. Y does not depend on Ws so this is easy:

$$\begin{aligned} \frac{dSE}{dW_i} &= \frac{d(\frac{1}{2}(Y - Y_o)^2\ )}{dW_i} \\ \frac{dSE}{dW_i} &= -\ (Y - Y_o)\ \frac{d(Y_o)}{dW_i} \end{aligned} \tag{4.5}$$

Since the squeezing function $\mathcal{L}$ is not linear, $d(Y_o)/dW_i$ is difficult to calculate. However, here for the sake of simplicity and readability, let's suppose that X and W are both small values. We are near the center of the sigmoid $\mathcal{L}$, where it is almost linear, so we can remove it. From equation 4.2, we will have:

$$Y_o = \Sigma_i X_i W_i + \theta \tag{4.6}$$

We derive only for the $\bar{W}$'s element W_i, the sum Σ_i goes away:

$$\frac{d(Y_o)}{dW_i} = X_i \tag{4.7}$$

so from (4.5), we will have:

$$\frac{dSE}{dW_i} = -(Y - Y_o)\ X_i \tag{4.8}$$

Here $(Y - Y_o)$ is the difference between expected value and the output of the perceptron, so we can simply call it the error $\varepsilon = (Y - Y_o)$. Now, from (4.4) we can calculate how to update our weights $\bar{W}$:

$$W_i = W_i + \alpha\varepsilon\ X_i \tag{4.9}$$

If we train the perceptron with examples and at each one we compute the error ε and update the weights as above equation (4.9), after a number of training loops the error will diminish and converge toward zero. The perceptron can recognize the input pattern and classifies it.

You want to test this? Just type a program in your computer and run it. We suppose that the training data are four medical tests, each of them consists of five values (those could be anything you want: blood sugar contents,

$$\frac{dSE}{dW_i} = \frac{d(\frac{1}{2}(Y - Y_o)^2\ )}{dW_i}$$
$$\frac{dSE}{dW_i} = -\ (Y - Y_o)\ \frac{d(Y_o)}{dW_i} \tag{4.5}$$

式 4.5 において $d(Y_o)/dW_i$ を計算するのは難しいので，ここで活性化関数 $\mathcal{L}$ を使って考えてみよう．また，単純化のために X と W は，原点付近の小さな値を用いることでシグモイド関数を線形として扱うことにする．そうすると，Y_o は以下の式で表すことができる．

$$Y_o = \Sigma_i X_i W_i + \theta \tag{4.6}$$

目的関数 Y_o を各要素 W_i で微分することで以下の式を得ることができる．

$$\frac{d(Y_o)}{dW_i} = X_i \tag{4.7}$$

式 4.5 に代入すると

$$\frac{dSE}{dW_i} = -(Y - Y_o)\ X_i \tag{4.8}$$

となる．ここで $(Y - Y_o)$ は，パーセプトロンの出力と期待値の差を表す．また単に誤差 $\varepsilon = (Y - Y_o)$ と呼んでも良いだろう．ここで式 4.4 を用いて，結合荷重 $\bar{W}$ を更新する次式が得られる．

$$W_i = W_i + \alpha\varepsilon\ X_i \tag{4.9}$$

もし複数の教師データで式 4.9 に従って結合荷重を更新することで誤差 ε を計算すれば，学習回数を重ねるごとに誤差は小さくなり，やがて 0 に近づくだろう（図 4.2）．そして，パーセプトロンは入力のパターンを学習することで入力の分類に成功する．

実際に試してみよう．自分のパソコンにコードを打ち込んで，実装しよう．まず，教師データとして，それぞれ 5 つの項目をもつ 4 つのメディカルテストを用意しよう．これは，測定したいものであれば何でも構わない．例えば血糖値や尿酸値等で，そのメディカルテストにおいて，想定される病気に対しての診断結果を 0 か 1 で評価することを目的とする．4 つの試験のうち 1 つを無

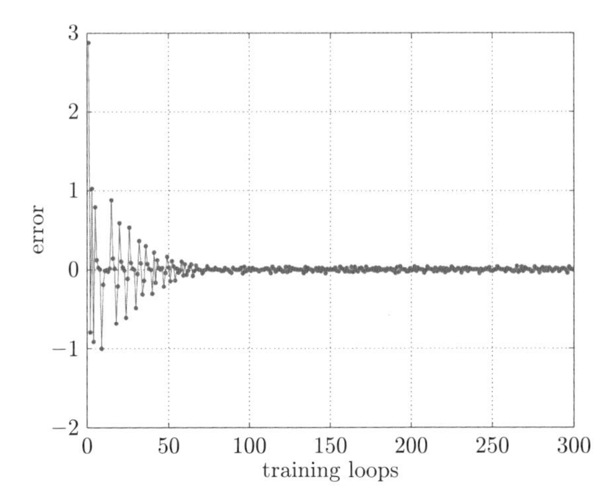

Fig. 4.2 The error output from a simulation of the simple perceptron in figure 4.1. The perceptron is trained with 4 examples of imaginary medical tests that can be positive or negative ("1" or "0"). The weights are updated with a gradient descent optimizer.

urine values etc.) The value "0" or "1" is our target, means that the results of medical test is positive or negative for a supposed disease. To train the perceptron we do $n = 300$ loops where we choose at random one of these four tests, over and over again. For each loop we calculate the perception output Y_o and update the weights like in equation (4.9).

Listing 4.1 Example in Octave

```
clear
close all
clf
training_data =...
[[0.5 0.3  0.5 1.1 0.95];...
[0.51 0.3  0.4 1.0 0.9];...
[0.49 0.9 0.8 0.9 0.5];...
[0.48 0.9  0.82 0.88 0.55]]
expected_y=[0 0 1 1];
err=zeros(0)

n=size(training_data)(1); % number of examples
```

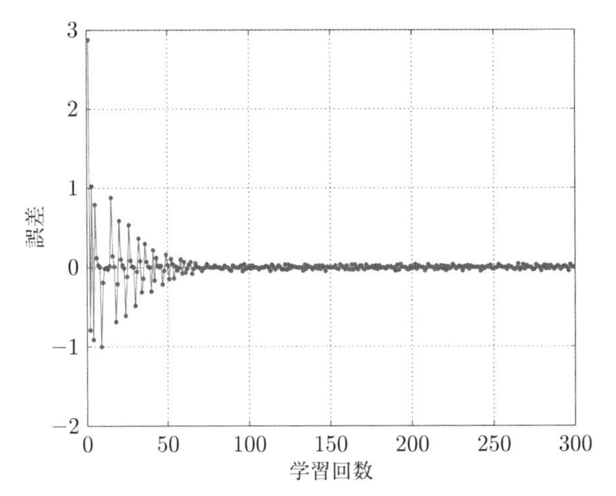

図 4.2 図 4.1 の単一パーセプトロンにおける誤差のシミュレーション結果．4 つの仮想的なメディカルテストを教師データとして学習したパーセプトロンは，最急降下法を用いた良否（1 か 0）判定器として機能する．

作為に選び $n = 300$ 回パーセプトロンを学習させる．もちろんその都度出力 Y_o を計算し，式 4.9 に従って結合荷重を更新する．

リスト 4.1 Octave サンプルコード

```
clear
close all
clf
training_data =...
[[0.5 0.3  0.5 1.1 0.95];...
[0.51 0.3  0.4 1.0 0.9];...
[0.49 0.9 0.8 0.9 0.5];...
[0.48 0.9  0.82 0.88 0.55]]
expected_y=[0 0 1 1];
err=zeros(0)

n=size(training_data)(1); % number of examples
```

```
i=size(training_data)(2); % number of inputs

W=2*rand(1,i)-1; % initial random weights
alpha=0.3 % descent speed
n=300% training loops
k=0; % counter
while k<n
      k=k+1;
      idx=randi(i-1); % choose one example
      X=(training_data(idx,:)); % get the input
      expected=expected_y(idx);
      result=dot(W,X);
      error=expected-result;
      err=[err error]; % append the value for plotting
      W = W + alpha*error*X; % update the W vector
end
plot(err,'.-')
ylabel("error")
xlabel("training loops")
```

The programming language we used is *Octave*. You can get it for free online, so you have no excuses not to verify by yourself that the network works! Notice that, in this example, we considered the bias $\theta = 0$ for simplicity. Also, the squeezing function $\mathcal{L}$ is not necessary, the network converges even without it. You can put this code in your PC, if you plot the error you will see that it will quickly go down to almost zero like in figure 4.2. This is fantastic: we just repeated over and over again the same 4 examples!

When the W are good, we can say the perceptron is *optimized* and the method used (the gradient descent in our case) is generally called the *optimizer*. In our case the square error (SE) was the cost function. This system now enables us to analyze a new, never known input and gives us an output that is reasonable. If the training data were obtained by real medical tests, we could give this perceptron to a doctor, and he/she can use it to classify

```
i=size(training_data)(2); % number of inputs

W=2*rand(1,i)-1; % initial random weights
alpha=0.3 % descent speed
n=300% training loops
k=0; % counter
while k<n
      k=k+1;
      idx=randi(i-1); % choose one example
      X=(training_data(idx,:)); % get the input
      expected=expected_y(idx);
      result=dot(W,X);
      error=expected-result;
      err=[err error]; % append the value for plotting
      W = W + alpha*error*X; % update the W vector
end
plot(err,'.-')
ylabel("error")
xlabel("training loops")
```

プログラミング言語には，Octave を使用する．Octave はフリーソフトウェアとして利用できるので，学生諸君は自分自身で上記のサンプルコードを気兼ねなく実践できるだろう．また，今回は例題なのでバイアスは $\theta = 0$ とした．さらにスクイージング関数もサンプルコードには出てこない．活性化関数なしでも収束可能なので，もし皆さんの PC でサンプルコードを実装し，誤差をプロットすると，たちまち 0 付近に収束することが確認できるであろう．4 つの教師データを単に繰り返し使用しただけだというのに，驚くべきことである．

結合荷重がうまく調整できたとき，パーセプトロンが最適化されたと言える．また，今回のサンプルコードで使用した最急降下法は，一般に最適化関数として知られている．目的関数は 2 乗誤差を用いた．このサンプルコードを使用すれば，とても簡単に新しい検査結果を解析することができる．もし実際のメディカルテストの教師データがあれば，パーセプトロンは医師の代わりに私たちの検査結果を自動的に診断してくれるだろう．

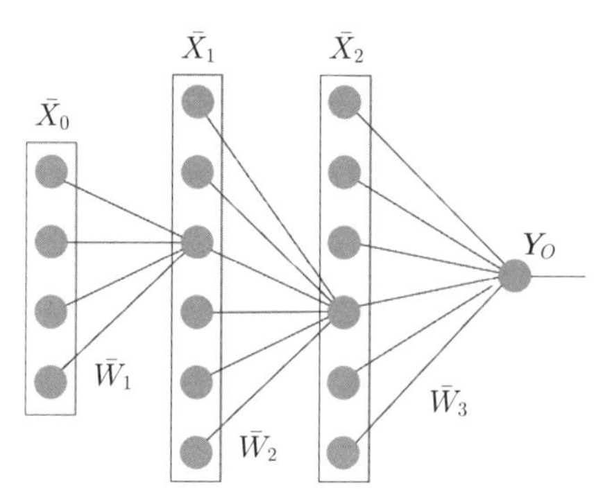

Fig. 4.3 A schematic representation of a multilayered artificial neural network. All neurons of one layer are connected to all the neurons of the previous layers (for clarity only one example of connections per layer is shown).

new data for positive and negative cases, automatically!

4.2 Artificial neural networks (ANN)

Now, instead of using only one neuron to make a perceptron, we can make a *network* of neurons, just like in our brain! A simple layered structure of neurons is sketched in figure 4.3. Here the $\bar{X}_0$ vector is the input, and the other circles are neurons arranged in *layers*.

We call the first layer of neurons $\bar{X}_1 = X_{11}, X_{12}, X_{13}...$, and the other $\bar{X}_2 = X_{21}, X_{22}, X_{23}...$, notice that these are vectors, not matrix. Every neuron of each layer is connected to all the neurons in the previous layer (for reasons of clarity, in figure 4.3 only few connections are shown).

In a structure like this, each neuron is the perceptron studied above. So it can *specialize* in recognizing a certain value pattern in its input. For example, assume that the first perceptron in the first layer (X_{11}) could specialize in recognizing positive slopes in the $\bar{X}_0$ input pattern, the second one (X_{12}) could recognize negative ones, and the third, could respond to positive concavity and the next to negative. You name a type of pattern in the input (a *feature* of $\bar{X}_0$) and you could imagine one of those perceptrons learning such input property! In the second layer each perceptron could

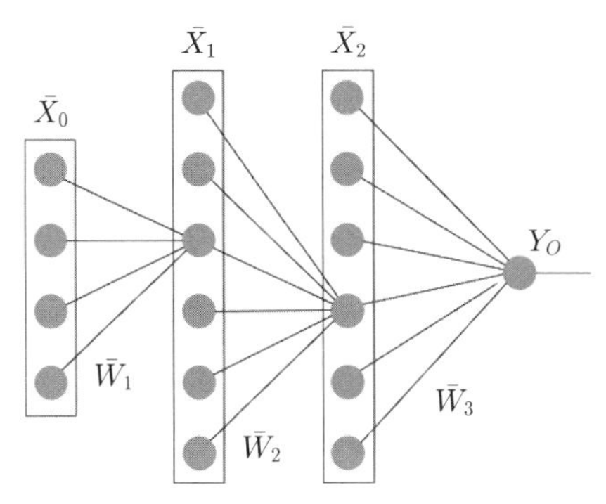

図 4.3　多層の人工ニューラルネットワークを示す模式図．ある層のすべてのニューロンは，その前の層のすべてのニューロンに接続されている（簡潔化のため，上図では 1 つの層ごとに一部の接続のみを示す）．

4.2　人工ニューラルネットワーク (ANN)

前節では 1 つのニューロンによってパーセプトロンを作ることに成功した．次に，ここでは，私たちの脳のように，複数のニューロンのネットワークを形成しよう．単一の層からなるニューロンの構造を図 4.3 に示した．入力 $\bar{X}_0$ ベクトルと各層のニューロンを図中に丸で示す．最初の層のニューロンを $\bar{X}_1 = X_{11}, X_{12}, X_{13}...$ とし，次の層のニューロンを $\bar{X}_2 = X_{21}, X_{22}, X_{23}...$ というようにしてベクトルで表記する．これらは行列ではなく層ごとにベクトルで表記する．各層の各ニューロンは 1 つ前の層のすべてのニューロンに接続される（簡潔化のために，図 4.3 では限られた接続しか示していない）．

この構造において各ニューロンはすでに説明したパーセプトロンとして扱われる．つまり，ある特定の入力パターンを認識することに特化することが可能である．例えば，最初の層の最初のパーセプトロン X_{11} は，入力 $\bar{X}_0$ のパターンが上向きと判断しているとする．さらにその次の X_{12} が下向きだとする．3 番目はまた上向き，次は下向きと上下を繰り返しているとする．すると入力のパターンを認識することができる．そして，入力パターンの特徴を抽出し，次第に各パーセプトロンが学習を繰り返し 1 つになっていく様子が想像できるだろう．第 2 の層の各パーセプトロンは特徴の組み合わせの認識に特化す

specialize in recognizing particular combinations of features. This neural network could learn much more complex features and each layer realizes a hierarchy of them. We have the potential to make a really smart machine here...!

Again, the point is to train the network with a set of examples in order to optimize the connection weights. How can we do that? This time $\bar{W}_1$ is not a one dimensional vector of values like in the case of the perceptron, but it is a rectangular matrix of dimension m and n. Here m is the number of inputs in $\bar{X}_0$ and n is the number of neurons in $\bar{X}_1$. The output Y_o will be calculated by applying equation (4.2) to all the neurons in the network. The calculation looks exactly the same as before but instead of the vector $\bar{W}_i$ we have rectangular matrices. For the first layer, we can write:

$$\bar{X}_1 = \mathcal{L}(\bar{X}_0 \cdot \bar{W}_1) \tag{4.10}$$

The neurons of $\bar{X}_1$ are connected to the second layer, so we repeat exactly the same calculation one layer forward:

$$\bar{X}_2 = \mathcal{L}(\bar{X}_1 \cdot \bar{W}_2) \tag{4.11}$$

and the third layer feeds the single neuron at the exit, the result will be:

$$\bar{Y}_o = \mathcal{L}(\bar{X}_2 \cdot \bar{W}_3) \tag{4.12}$$

Notice that we neglected again the bias θ for the sake of simplicity. The process of obtaining the output Y_o from the input $\bar{X}_0$ is called *feed forward* propagation.

Now that we know the output Y_o we can calculate the error by the following:

$$\bar{\varepsilon} = Y - Y_0 \tag{4.13}$$

We can immediately update the weights near the output W_3 using an equation similar to (4.9)

ることが可能である．ニューラルネットワークでは，パーセプトロンが階層構造をもつことで，複雑な予測を学習することができる．とても賢いコンピューターを実現できそうである．

ネットワークが学習するということは，複数の教師データを用いて結合荷重を最適化することであったことを思い出してほしい．この場合どのようにすれば良いのだろうか．パーセプトロンのときと違って，$\bar{X}_1$ は 1 次元のベクトルデータではなく，2 次元の m 行 n 列の行列である．ここで m は $\bar{X}_0$ の入力の数であり，n は $\bar{X}_1$ のニューロンの数である．また出力 Y_o はネットワーク中のすべてのニューロンに式 4.2 を適用させることで導かれる．この計算を行うことは，単に式 4.2 において結合荷重をベクトル $\bar{W}_i$ で置き換えているように見えるが，ベクトル $\bar{W}_i$ の代わりに行列を用いているのである．最初の層に関して次のように記述する．

$$\bar{X}_1 = \mathcal{L}(\bar{X}_0 \cdot \bar{W}_1) \tag{4.10}$$

$\bar{X}_1$ で表されるニューロンは第 2 層目のニューロンと結合する．そして，次々に各層を計算していく．

$$\bar{X}_2 = \mathcal{L}(\bar{X}_1 \cdot \bar{W}_2) \tag{4.11}$$

第 3 層目は単一のニューロンに接続し，結果は次式になる．

$$\bar{Y}_o = \mathcal{L}(\bar{X}_2 \cdot \bar{W}_3) \tag{4.12}$$

ここでも問題の簡潔化のためにバイアスは無視した．入力 $\bar{X}_0$ から出力 Y_o を得る過程をフィードフォワード伝播と呼ぶ．

出力 Y_0 を利用して，誤差を計算することができる．

$$\bar{\varepsilon} = Y - Y_0 \tag{4.13}$$

式 4.9 にならって結合荷重 $\bar{W}_3$ を更新する．

$$\bar{W}_3 = \bar{W}_3 + \alpha \bar{X}_2^T \cdot \bar{\varepsilon} \tag{4.14}$$

T は，行列の転置の操作を意味する．パーセプトロンに関する式 4.9 を思い出

$$\bar{W}_3 = \bar{W}_3 + \alpha \bar{X}_2^T \cdot \bar{\varepsilon} \tag{4.14}$$

where the apex T symbolize the transpose of the matrix. Again, this is equation 4.9 of the perceptron case with the difference that $\bar{X}$ and $\bar{W}$ are now rectangular arrays and not vectors. For the previous layer we can apply the same equation, but we need to calculate it from the output using the reversed dot product, so we have:

$$\begin{aligned} \varepsilon_2 &= \bar{\varepsilon} \cdot \bar{W}_3^T \\ \bar{W}_2 &= \bar{W}_2 + \alpha \bar{X}_1^T \cdot \bar{\varepsilon}_2 \end{aligned} \tag{4.15}$$

and we do the same thing similarly for the first layer:

$$\begin{aligned} \varepsilon_1 &= \bar{\varepsilon} \cdot \bar{W}_2^T \\ \bar{W}_1 &= \bar{W}_1 + \alpha \bar{X}_0^T \cdot \bar{\varepsilon}_1 \end{aligned} \tag{4.16}$$

This is the way to update the Ws of all the layers of the network. The procedure can be generalized, the input X_0 or the output Y_o do not have to be simple vectors, but could be multidimensional matrices.

Now, let's try again in practice. Let's use exactly the same training data we used for the perceptron. This time the network has five neurons of input, two hidden layers of ten neurons each, and one single output. Use this program

Listing 4.2 Example in Octave

```
clear
close all
clf
function out=sqz(val)
     out=atan(val);
end
training_data =...
[[0.5 0.3  0.5 1.1 0.95];...
[0.51 0.3  0.4 1.0 0.9];...
```

してみよう．$\bar{X}$ と $\bar{W}$ をベクトルではなく行列として扱ってみよう．前の層に誤差を伝搬するには，逆行列の積を用いて計算することで出力を得ることができる．

$$\begin{aligned} \varepsilon_2 &= \bar{\varepsilon} \cdot \bar{W}_3^T \\ \bar{W}_2 &= \bar{W}_2 + \alpha \bar{X}_1^T \cdot \bar{\varepsilon}_2 \end{aligned} \tag{4.15}$$

そして第 1 層に関しても同様に修正し，更新する．

$$\begin{aligned} \varepsilon_1 &= \bar{\varepsilon} \cdot \bar{W}_2^T \\ \bar{W}_1 &= \bar{W}_1 + \alpha \bar{X}_0^T \cdot \bar{\varepsilon}_1 \end{aligned} \tag{4.16}$$

この方法により，ネットワークすべての層の結合荷重 W を更新することができる．一般に入力 X_0 や出力 Y_o はベクトルだけでなく多次元行列に拡張することができる．

それでは，再びサンプルコードを動かしてみよう．教師データはパーセプトロンのときに使用したものとまったく同じデータで試してみよう．このモデルでは，入力層として 5 個のニューロン，2 つの隠れ層として各々 10 個のニューロン，そして 1 つの出力をもつ．それでは，サンプルコードを実装してみよう！

リスト 4.2　Octave サンプルコード

```
clear
close all
clf
function out=sqz(val)
      out=atan(val);
end
training_data =...
[[0.5 0.3  0.5 1.1 0.95];...
[0.51 0.3  0.4 1.0 0.9];...
```

```
[0.49 0.9 0.8 0.9 0.5];...
[0.48 0.9  0.82 0.88 0.55]]
expected_y=[0 0 1 1];
n=size(training_data)(1); % number of examples
i=size(training_data)(2); % number of inputs
nh1=10;
nh2=10;
W1=2*rand(i,nh1)-1; % initial random weights
W2=2*rand(nh1,nh2)-1;
W3=2*rand(nh2,1)-1;
alpha=0.3 % descent speed
n=300% training loops
err=zeros(0)
k=0; % counter
while k<n
 k=k+1;
 idx=randi(i-1); % choose one example
 X0=(training_data(idx,:)); % get the input
% feed forward
 X1=sqz(X0*W1);
 X2=sqz(X1*W2);
 result=sqz(X2*W3);
% back propagate the error
 expected=expected_y(idx); % get the value
 error=expected-result; % error
 dX2=((error)*transpose(W3)); % back error
 dX1=((dX2)*transpose(W2)); % back error
 % update the coefficients W
 W3 = W3 + alpha*transpose(X2)*error; % update the W vector
 W2 = W2 + alpha*transpose(X1)*dX2; % update the W vector
 W1 = W1 + alpha*transpose(X0)*dX1; % update the W vector
 % memorize the error for plotting later
 err=[err error]; % append
end
plot(err,'.')
ylabel("error")
xlabel("training loops")
```

```
[0.49 0.9 0.8 0.9 0.5];...
[0.48 0.9  0.82 0.88 0.55]]
expected_y=[0 0 1 1];
n=size(training_data)(1); % number of examples
i=size(training_data)(2); % number of inputs
nh1=10;
nh2=10;
W1=2*rand(i,nh1)-1; % initial random weights
W2=2*rand(nh1,nh2)-1;
W3=2*rand(nh2,1)-1;
alpha=0.3 % descent speed
n=300% training loops
err=zeros(0)
k=0; % counter
while k<n
 k=k+1;
 idx=randi(i-1); % choose one example
 X0=(training_data(idx,:)); % get the input
% feed forward
 X1=sqz(X0*W1);
 X2=sqz(X1*W2);
 result=sqz(X2*W3);
% back propagate the error
 expected=expected_y(idx); % get the value
 error=expected-result; % error
 dX2=((error)*transpose(W3)); % back error
 dX1=((dX2)*transpose(W2)); % back error
 % update the coefficients W
 W3 = W3 + alpha*transpose(X2)*error; % update the W vector
 W2 = W2 + alpha*transpose(X1)*dX2; % update the W vector
 W1 = W1 + alpha*transpose(X0)*dX1; % update the W vector
 % memorize the error for plotting later
 err=[err error]; % append
end
plot(err,'.')
ylabel("error")
xlabel("training loops")
```

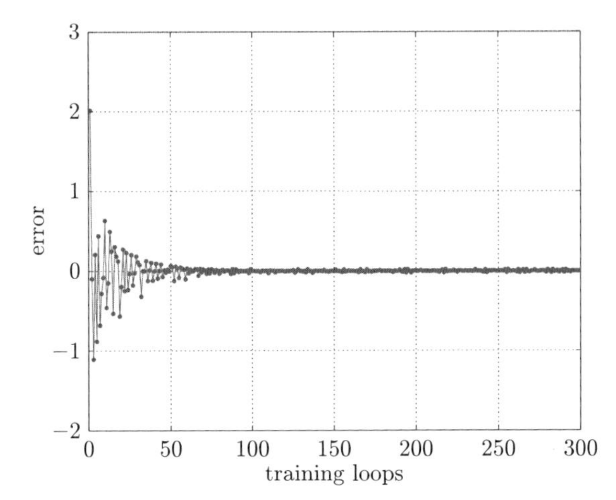

Fig. 4.4 The output of the artificial neural network of three layers described. The gradient descent method works and makes the error converge soon. If you do different tests, you will notice that every time the descent looks different. In some cases, the error oscillates more violently, whereas in other cases, like this one, goes to zero more quietly.

Try to run this program and you will obtain an output similar to figure 4.4. Notice the function sqz, that was not necessary for the perceptron, is fundamental now. It corresponds to the squeezing function $\mathcal{L}$ described above. Try to remove it and the network will not converge!

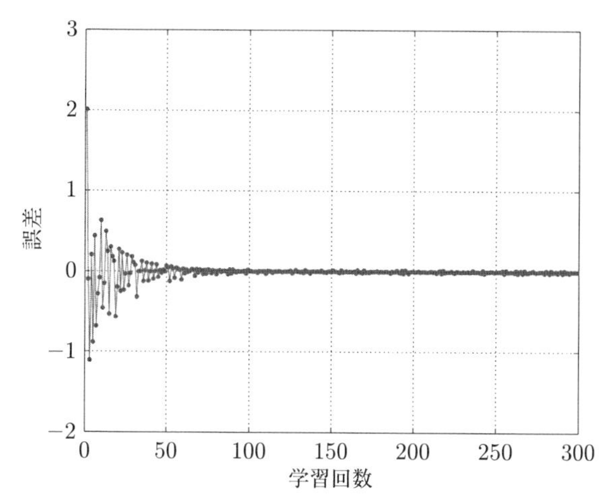

図 4.4　3 層の人工ニューラルネットワークの出力例．最急降下法によって，誤差はすぐに収束する．もし新たなデータでプログラムを実装する場合，その振る舞いは毎回異なるだろう．例えば，誤差が発振してしまったり，一方では，より 0 付近で収束したりする．

実際にプログラムを実行したら，図 4.4 のような結果が得られただろう．サンプルコード内の sqz に着目してみよう．基本的には，単一のパーセプトロンには必要ないが，ここでは必須となる．sqz は活性化関数 $\mathcal{L}$ に対応する．sqz をサンプルコード内から削除してみよう．そうすれば計算が収束しない様子が確認できるだろう！

Think about these questions

- What is the biological meaning of the *squeezing* function?
- What is a *perceptron*?
- What is the *cost function*?
- What does the *gradient descent* method do?
- What is the function of the weights W in a perceptron?
- If we connect many perceptrons together, what do we obtain?
- How does the gradient descent method change if we have many layers?
- Can a network be optimized even without bias or squeezing function?

考えてみよう

- 生物学における活性化関数の意味はどのようなことか？
- パーセプトロンとはどのようなものか？
- 誤差関数とはどのようなものか？
- 最急降下法を説明せよ．
- パーセプトロンにおいて結合荷重の役割はどのようなものか？
- 多くのパーセプトロンが互いに結合したとき，どのような結果が予測されるか？
- 多層のニューラルネットワークにおける最急降下法はどのように書き換えれば良いのか？
- バイアスや活性化関数なしでネットワークを最適化することは可能か？

Chapter **5**

Deep learning

There is a generic term called *deep learning.* This expression refers to the construction of artificial neural networks and to the process of teaching them how to operate and solve problems. Similar to what we did in chapter 4, we can teach a network to read medical data and understand if they correspond to positive or negative tests. Sometimes deep learning is used to indicate more complex and structured neural networks than the one in chapter 4. Hereafter, we will introduce the most basic networks used in deep learning.

A convolutional neural network (CNN) is a neural network in which specific mathematical calculations are performed. In conventional ANNs the input of a neuron is the multiplication of the weights W with the previous layer outputs. In a CNN, before to these operation, the input is processed to extract meaningful information. This is done with particular functions called *convolutions.* In addition to that, other mathematical processes are done to simplify and reduce the size of the network. After this pre-processing, the data are fed to a conventional neural network for the final optimization of W. We are going to describe schematically these operations.

First of all, the scheme of a standard CNN is as in figure 5.1. Let's suppose that the input vector of our neural network is a gray-scale image. Also, let's think that our goal is to classify these images, that is to understand what they represent. For example, the input images could be handwritten numbers from 0 to 9. As a practical example we are going to use the MNIST database[NIST and Technology, 1995]. Our images are exactly 28×28 pixels

第 5 章

ディープラーニング

この章ではディープラーニングについて取り上げる[1]．ディープラーニングは，人工ニューラルネットワークにより構成され，学習により問題を解決することができる．例えば，すでに第 4 章で学んだように，ある病気に関する良否の判断を，いくつかのメディカルデータによってそのネットワークに学習させることができる．しかし，一般的にディープラーニングは，ニューラルネットワークよりも複雑な構造をもっている．そこで，ここではディープラーニングの特に重要な部分を説明する．

畳み込みニューラルネットワーク (convolutional neural networks: CNN) は，特別な数学的処理を用いたニューラルネットワークである．従来の人工ニューラルネットワークでは，ニューロンの入力は，重み W と前に計算された出力の乗算の結果をニューロンの入力として利用する．一方，CNN では，それらの処理の前に，入力からより有力な情報を得るための工夫を行う．この処理は畳み込みと呼ばれる特定の関数によって行われる．また CNN では，ネットワークのサイズを小さくするような他の単純な数学的処理も行う．これらの処理を前段階に行った後，データは従来のニューラルネットワークによって W を最適化する．それらの演算がどのようなものか，概略を次に示していく．

図 5.1 には，標準的な CNN の概要を示した．まず入力にはグレースケールの画像データを仮定しよう．目的は入力された画像データをそれらのイメージごとに分類することである．つまり，それらの画像に何が表現されているかを理解させるということである．例えば入力データは手書きの 0 から 9 の数字である．MNIST（エムニスト）と呼ばれるデータベースを用いることでサンプルデータを手に入れることが可能である [NIST and Technology, 1995].

1) ディープラーニング（深層学習）：deep learning.

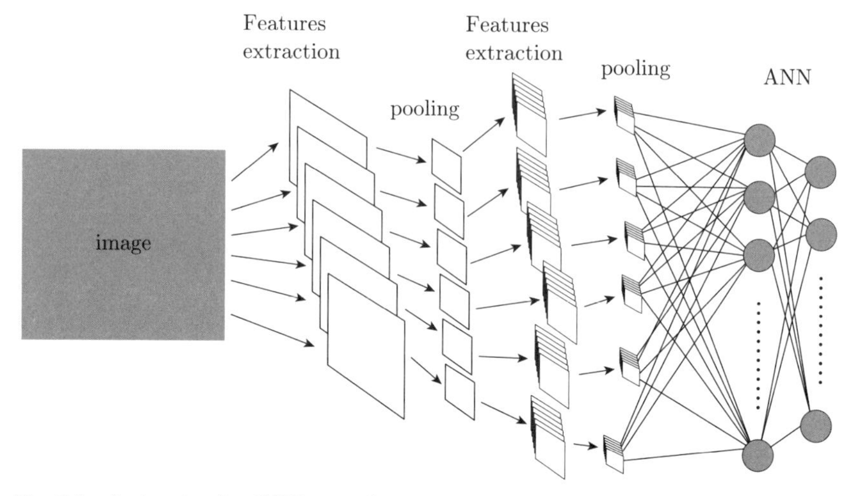

Fig. 5.1 A sketch of a CNN neural network. An image, or any array of data, is passed through several filters by a convolution process, (equation 5.1). The new data created are reduced in size by pooling and rectified in value. This process can be repeated few times. The final result is fed to an artificial neural network that it is optimized for the desired output through back-propagation or other algorithms.

in order to be compatible with the database images. MNIST contains 60000 handwritten digits on a 28×28 grey-scale grid. So the input layer of our network will be an array of 784 values.

As we learned in chapter 4, in a conventional ANN network, each neuron of the second layer will receive an input that is the weighted sum of all the outputs of the previous layer. The weights are the connections strengths W_{ij}. In a CNN, the input vector is transformed through filters instead.

If we have n filters, the transformation will produce n new images. The filtering is the main characteristics of CNNs. As said above, this is done by *convolutions*, from which the CNN takes its name. For signals that depend from only one variable, the convolution is expressed by this formula:

$$(f * g)(t) = \int_{-\infty}^{\infty} g(\tau) f(t-\tau) d\tau \tag{5.1}$$

Where f and g are two functions of time (the two signals) and the symbol $(f * g)$ represents the resulting convolution function. As you can see from

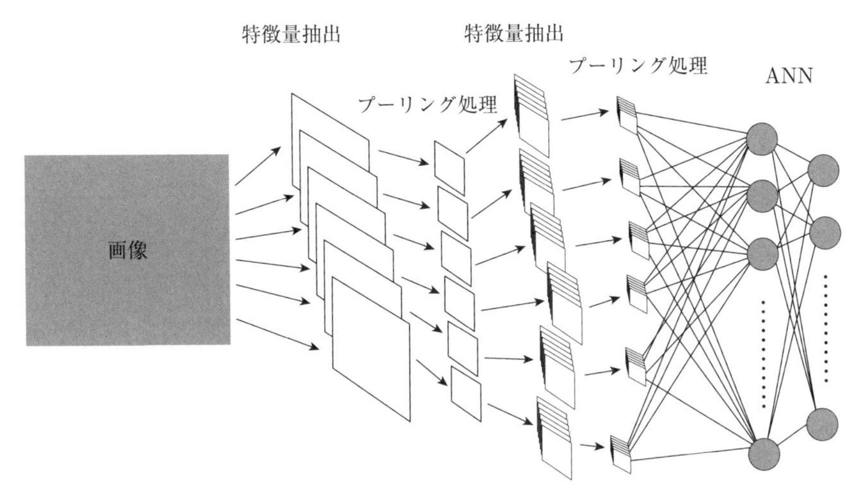

図 5.1　CNN の概略図．画像データは畳み込み処理によるいくつかのフィルターを通る（式 5.1）．フィルター通過後の新しいデータは値のプーリングや正規化によってデータサイズが圧縮される．この処理を数回繰り返した出力データを人工ニューラルネットワークに渡す．人工ニューラルネットワークは，誤差逆伝搬法や，その他のアルゴリズムによって最適化される．

MNIST には 28×28 ピクセルの手書きの数字のグレースケールの画像データが 6 万件登録されている．今のネットワークの例では入力層として配列数 784 の配列が与えられる．第 4 章で従来のニューラルネットワークを学んだときは，第 2 層目の各ニューロンは，それらに接続される結合荷重 W_{ij} すべてを足し合わせた．一方，CNN の入力ベクトルはいくつかのフィルターによって変換される．

もし N 個のフィルターを用いれば，N 個の新たな画像を得る．それらのフィルターは CNN の最大の特徴である．CNN の名前の通り，畳み込みと呼ばれる処理が用いられる．1 次元における畳み込みの数学的表現は以下の式になる．

$$(f * g)(t) = \int_{-\infty}^{\infty} g(\tau) f(t-\tau) d\tau \tag{5.1}$$

ここで，$f * g$ は，f と g の畳み込み積分を表す．畳み込みは 2 つの関数間のフィルターとして表現される．式 5.1 からも畳み込みが 2 つの関数の掛け算を意味していることがわかる．その関数の 1 つは信号 $f(t-\tau)$ で時間に依存

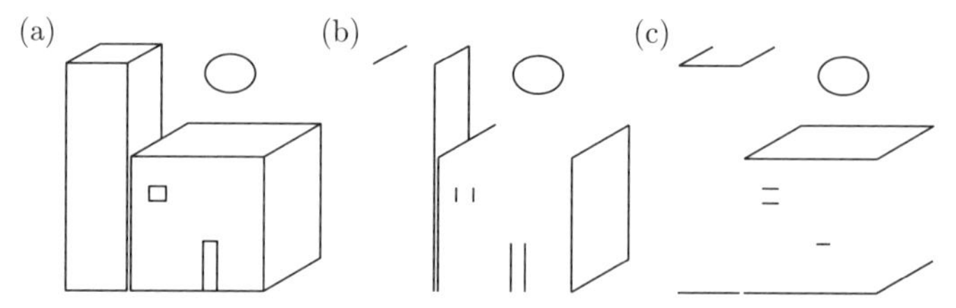

Fig. 5.2 Example of feature extraction: an original image (a) is filtered for vertical (b) and horizontal (c) lines. In a CNN there are many types of convolutions that extract lines of various inclinations, filter our colors, corners or other geometrical features.

equation 5.1, convolution means, in a sense, the *multiplication* of two functions. However, one of the function is changing with time independently, this is the signal $f(t-\tau)$ of equation 5.1, whereas the other function g that does not depend on time is the filter. You can find an intuitive visualization of the convolution in a nice applet made by a student of the university of Brno[Kaňok, 2018].

Once you understand the meaning of convolution in one dimension, you can easily extend the idea of filter in two dimensions or more. In two dimensions, the variable t above could be instead an image Σ, and the filter would be a smaller array $g(\sigma)$ that acts on the image. Mathematically it can be expressed by modifying equation 5.1 as the following:

$$(f*g)(\Sigma) = \iint_{\sigma} g(\sigma) f(\Sigma - \sigma) d(\sigma) \tag{5.2}$$

Here the integral is over the small filter surface σ, instead Σ is the larger surface that represents the image.

The convolution could realize any filtering that add useful meaning to the image. For example, the filter could recognize edges oriented at a particular angle. All details of the image that are oriented to that angle will be emphasized, the other details will be reduced. So, from our starting vector Σ (the image), the network could produce n new images Σ_n, each corresponding to the different σ_n filtering. Like in the example of figure 5.2. Filters could detect edges oriented in different angles, but we can make filters that

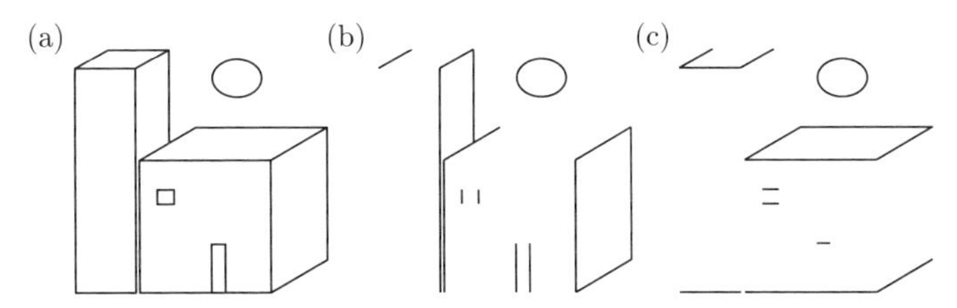

図 5.2 特徴量抽出の例：元画像 (a) は，フィルターによって縦線画像 (b) と横線画像 (c) に処理する．CNN には多種の畳み込みフィルターがあり，線や傾斜，色やコーナー，その他の幾何学的な特徴を抽出可能である．

するが，もう一方の関数であるフィルター g は時間に依存しない．ウェブでも畳み込み積分の多くのアプリケーションを見つけることができ，チェコのブルーノ大学で知られる良いアプリケーションがある [Kaňok, 2018].

1 次元の畳み込みを理解することができれば，2 次元，3 次元とそれ以上の次元に関してフィルターの概念を拡張するのは容易である．式 5.1 を 2 次元に拡張すると，時間を入力画像データと置き換え，より小さな配列を作用して処理することを考える．以下に式 5.1 を修正した 2 次元の畳み込み積分の数学的表現を示す．

$$(f * g)(\Sigma) = \iint_{\sigma} g(\sigma) f(\Sigma - \sigma) d(\sigma) \tag{5.2}$$

ここで Σ は入力画像であり，$g(\sigma)$ はフィルターである．σ は Σ より小さい画像サイズである．

畳み込みの処理はいくつかのフィルターによって画像データをより有効に利用できる．例えば，エッジフィルターによる処理などがある．フィルターによって画像の特徴量を際立たせることで角や曲線を強調することができる．ここで 1 つの入力画像 Σ を入力するとネットワーク内の σ_n によって n 個の新たな画像 Σ_n を得ることができる．図 5.2 には，模式図を示す．いくつかのフィルターを用いて異なる角度を検出することができる．フィルターによって細部の異なる角度を検出することで角や曲線を強調することができる．フィルター後の新しい画像は，一定の特徴を強調する．この処理工程は特徴抽出法と呼ばれている．さらにその後の処理として正規化処理を行う．正規化線形関数 (Rectified Linear Unit: ReLU) と呼ばれる処理が一般的で，画像データの各

extract colors or that emphasize corners or curves. Each of the new pictures enhances a certain *feature* of the image. This process is called *feature extraction*. After these filtering we have usually a normalization process in which the values are *rectified*. A common method to do so is called ReLU (Rectified Linear Unit). With it, the value of the pixel is set to zero if negative, and kept as it is if positive.

To reduce the number of data a process called *pooling* is used. There are many possible pooling methods. The most common is as simple as taking each four pixels in the image and substituting them with a single one corresponding to the maximum of the four. This produces a smaller image but hopefully maintains the great part of the features information in it. Mathematically, the processes of filtering, rectifying and pooling are all convolutions. Those processes can be repeated in successive layers. We will obtain more images, but smaller and smaller. If we choose the number of filters and the range of pooling well the amount of neurons in the network will still be reasonable. The final result will be fed to a conventional ANN. A learning procedure, similar to what studied in chapter 4 will be used to optimize the connection strengths W, then the CNN will learn.

The advantage of a CNN against a conventional ANN is that they can resolve classification problems much better and with more precision than ANNs. This is because the preliminary filtering extracts features that are relevant to the classification. For example, to recognize characters the curvature and the edge of the image is very important. A simple ANN doesn't know anything about edges or curvature, instead the convolution process (the filtering) enhance these features and feed them to the final ANN. This produces much better results than the sole ANN.

Also, please remember this important fact: in the optimization process any parameter of the network can be changed! Not only the connection strengths W, but also any other parameter. Even those in the filtering or pooling process. This means that we can keep free parameters in the filtering

ピクセルに対して値が負の場合は 0 を代入し，値が正であれば同じ値を代入するといった処理である．

その後，プーリングと呼ばれる処理を用いてデータ量を減らす．プーリング処理にはいくつかの種類が存在する．その中で最も単純で代表的な例は，データ内の 4 つの値をその中の最大値 1 つに置き換える方法である．オリジナルの画像に対してそれぞれ特徴量を抽出したデータができる．さらに，これらの N 個のデータに対して，必要に応じて正規化やプーリングの処理を繰り返し行うことができる．そうすれば，より小さなサイズのデータをより多く得ることができる．これらのフィルターを上手に選択することで，各層のニューロンの数を調整することができる．学習の手順は第 4 章と同様で，結合荷重 W を最適化する．これにより CNN は学習を行う．

従来の ANN（人工ニューラルネットワーク）に比べて CNN の利点は，その処理能力にある．CNN は，ANN に比べて分類問題をより早く，より正確に解決することができる．それは，分類問題に対して適したフィルター処理を行うことで実現する．CNN に画像の特徴を認識させるためには，画像のカーブやエッジの情報はとても重要である．畳み込み等のフィルター処理を用いて，特徴量を際立たせることで，その後の ANN の処理において，より精度の高い結果を得ることができる．

また，次の重要なことを思い出して欲しい．最適なネットワークにするためにはいくつかのパラメーターを最適化する必要があった．それは結合荷重 W だけではなかった．畳み込みによるフィルターやプーリング処理に関するパラメーターも最適化する必要がある．つまり，各々のフィルター処理のいくつかのパラメーターを設定し，それらを分類問題に対して最適化していく．設定するパラメーターの例としては，ふちの角度や曲線の曲率，またはカラーフィルターにおける RGB の値などがある．

(for example the edge angles, the curvature, or RGB values of color filtering or anything else) and the optimizer will choose for us the best values during the training.

Let's try to make a CNN in practice. Instead of building each neuronal connection by ourselves, we use a commands library that already defines the network structure. There are many of these libraries available, here we choose the one called *Keras* [Chollet *et al.*, 2015]. With the help of the library, we need only a brief program in python language[1] to realize the network. In the first lines of the program we import the libraries

```
from numpy import *
import keras
from keras.datasets import mnist
from keras.models import Sequential
from keras.layers import Dense, Flatten, Conv2D, MaxPooling2D
from keras.utils.np_utils import to_categorical
```

then we set commands to read the MNIST dataset:

```
(Xtrain, Otrain), (Xtest, Otest) = mnist.load_data()
```

This is composed by thousands of 28×28 pixels grey-scale images representing hand written characters, the ten digits between 0 and 9 (see figure 5.3).

The dataset is divided in 60000 images used for the training of the network (Xtrain in this example) and 10000 used for testing it (Xtest). For each image in Xtrain there is the corresponding Otrain value, that is the desired network output. This value represents the correct classification of the image: an integer number between 0 and 9. The same scheme applies to the images Xtest and their corresponding classification values Otest. It is necessary for the Keras library that the database values are in a conventional shape.

1) You need to install python on your computer. Also you need the python libraries "numpy", "keras" and "tensorflow". To install those libraries, you can use commands like `pip install keras` or similar commands, depending on your OS and computer type.

次に，CNN の実装を行ってみよう．ここでは，自身でニューラルネットワークを構築する代わりに，Keras というライブラリーを使って実装を試みよう [Chollet *et al.*, 2015]．Keras は 2015 年製で，コードを簡潔にまとめるためのライブラリーが Python 言語で用意されている[2]．次の最初の数行でライブラリーをインポートする．

```
from numpy import *
import keras
from keras.datasets import mnist
from keras.models import Sequential
from keras.layers import Dense, Flatten, Conv2D, MaxPooling2D
from keras.utils.np_utils import to_categorical
```

下記は，MNIST のデータを読み込むコマンドである．

```
(Xtrain, Otrain), (Xtest, Otest) = mnist.load_data()
```

ここで 0 から 9 までの数字を手書きした 28×28 ピクセルのグレースケール画像を大量に用意する（図 5.3）．

それらのデータは，ネットワーク学習をさせるための 6 万枚の画像（この例では Xtrain）と，学習後のテスト用の 1 万枚の画像（この例では Xtest）である．Xtrain の各画像はネットワークの出力として Otrain に変換される．Otrain の値によって画像が 0 から 9 のどの整数なのか正しく分類される．同様な形式で Xtest も Otest に変換され，正しく分類される．Keras ライブラリーには教師データもある．入力画像はテンソルの形式 $(\mathrm{T}(\mathrm{L}, x, y, 1))$ でなければならない．ここで L は学習枚数に相当し，サンプルコードの例では 6 万枚である．x および y の値は画像サイズでこの例では 28×28 ピクセルである．1 は色の値の要素であり，ここでは，グレースケールを使用しているので

2) Python をインストールしてみよう．また，同時に “Numpy” や “Keras”，“Tensorflow” といったライブラリーもインストールしてみよう．ライブラリーのインストールは，`pip install keras` のようなコマンドを用い自身の OS に合わせて行おう．

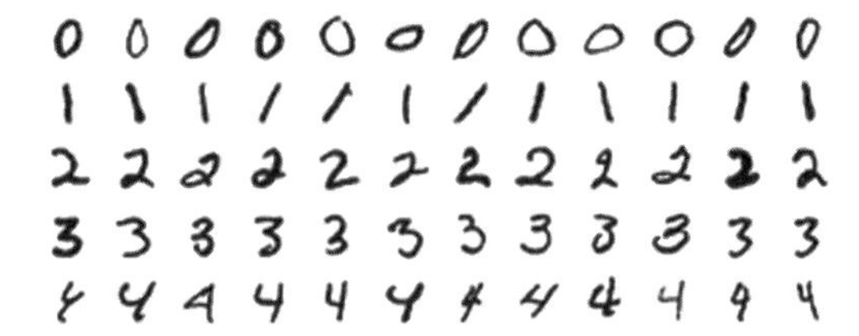

Fig. 5.3 Example characters from the MNIST database.

The input image must be a tensor T(L,x,y,1), where L is the length of the training set, in our case 60000, x and y are the size in pixels of the image, in our case 28 and 28, and 1 is the color dimensions. In our case we have only grayscale, so this must be 1 (for a RGB image that would be 3). The following lines convert the input in the right shape and put it to the required numerical type ('float32'/255).

```
Xtrain = Xtrain.reshape(60000, 28,28,1)
Xtest = Xtest.reshape(10000, 28,28,1)
Xtrain = Xtrain.astype('float32') / 255
Xtest = Xtest.astype('float32') / 255
```

The last conversion required by the Keras library is the following: the 10 classes, must be converted in vectors of zeros and ones. Our classes are a number from 0 to 9, the 10 digits in MNIST database. In fact they are vectors of 10 elements. The elements are all zeros and a single one is placed in the position corresponding to the class. For example, if our first five elements of Otrain were the vector [3,2,8,0,5], the conversion would give out a bidimensional vector of this type

$$\begin{array}{c}[0,0,0,1,0,0,0,0,0,0]\\ [0,0,1,0,0,0,0,0,0,0]\\ [0,0,0,0,0,0,0,0,1,0]\\ [1,0,0,0,0,0,0,0,0,0]\\ [0,0,0,0,0,1,0,0,0,0]\end{array} \tag{5.3}$$

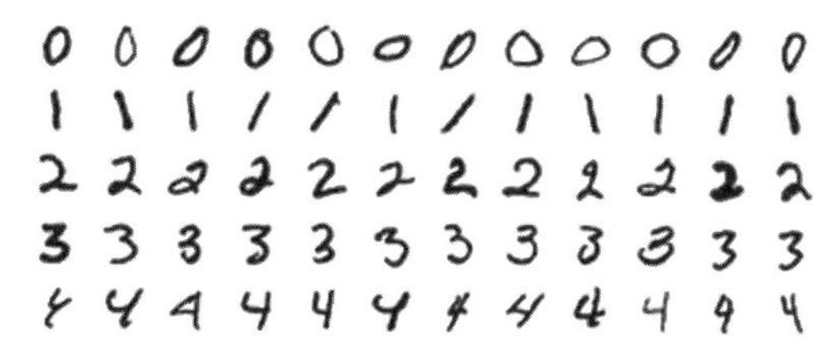

図 5.3　MNIST データベースの画像例

要素数は1であるが，RGB のカラーでは3になる．下記の数行では，データを正しい形式に変換し，必要な数値形式に置き換えている ('float32'/255).

```
Xtrain = Xtrain.reshape(60000, 28,28,1)
Xtest = Xtest.reshape(10000, 28,28,1)
Xtrain = Xtrain.astype('float32') / 255
Xtest = Xtest.astype('float32') / 255
```

Keras ライブラリーによって以下のように変換される．10 のクラスは0と1からなるベクトルに変換される．ここでいうクラスは MNIST データベースの0から9の数字を意味する．つまり，10 個の要素をもったベクトルである．また，そのベクトルは0と1の組み合わせによって0から9までの数を示す．例えば，Otrain の入力ベクトルが [3,2,8,0,5] のとき，変換後は次のような2次元の出力が得られるだろう．

$$
\begin{gathered}
[0,0,0,1,0,0,0,0,0,0] \\
[0,0,1,0,0,0,0,0,0,0] \\
[0,0,0,0,0,0,0,0,1,0] \\
[1,0,0,0,0,0,0,0,0,0] \\
[0,0,0,0,0,1,0,0,0,0]
\end{gathered}
\tag{5.3}
$$

この変換は次の関数によって行われる．

This conversion is done by the function:

```
to_categorical
```

So we add these two lines to do that:

```
Otrain = to_categorical(Otrain)
Otest = to_categorical(Otest)
```

Now the data are ready in the correct shape. We can construct our neural network with few simple commands:

```
model = Sequential()
model.add(Conv2D(32,kernel_size=(3,3),activation='relu',
input_shape=(28,28,1)))
```

This will create the first 32 feature maps with (3x3) filters. The activation function will be a *rectified linear unit*, called by Keras authors "relu". Then the resulting feature maps will be reduced in size by a pooling function that will select the maximum pixel among a (2,2) region. The library Keras calls this "MaxPooling2D":

```
model.add(MaxPooling2D(pool_size=(2, 2)))
```

Now the rest of the network is a conventional ANN, like we have studied in chapter 4. In Keras this is done with few lines:

```
model.add(Flatten())
model.add(Dense(128, activation='relu'))
model.add(Dense(num_classes, activation='softmax'))
```

The "Flatten()" element changes the shape of the input from the previous layer to a *flat* vector[2)]. The *Dense* element creates a standard ANN layer with 128 neurons. It is called dense, because each neuron of the layer is fully (or densely) connected with the previous input layer, as in chapter 4. Up to now, the code just set up structural definitions and parameter for the

2) a flat vector is the result of the conversion of a multidimensional array to a one dimensional vector.

```
to_categorical
```

この関数により，たった2行で変換のコードを記述することができる．

```
Otrain = to_categorical(Otrain)
Otest = to_categorical(Otest)
```

この関数を使用するために，入力の形式を適切にする必要がある．しかし，これらの準備にも数行のコードでニューラルネットワークの構成が実現できる．

```
model = Sequential()
model.add(Conv2D(32,kernel_size=(3,3),activation='relu',
input_shape=(28,28,1)))
```

これは 3×3 のフィルターを用いて，32の特徴量をマップした出力を得る．活性化関数は正規化関数である．Kerasでは "relu" と表される．その後，2×2 のプーリングフィルターを用いて最大値を抜き出し，データサイズを減少させる．Kerasでは "MaxPooling2D" と表される．

```
model.add(MaxPooling2D(pool_size=(2, 2)))
```

残りは，第4章で学んできた従来のANNの処理を実装する．これもKerasでは数行で記述が可能である．

```
model.add(Flatten())
model.add(Dense(128, activation='relu'))
model.add(Dense(num_classes, activation='softmax'))
```

"Flatten()" 関数は1つ前の層からの入力をフラットベクトルに変換する[3)]．またDenseの関数により，1層が128個のニューロンからなるネットワークを構成する．第4章で学んだように，隣り合う各層ですべてのニューロンがその前の層のニューロンに接続されていて，これまで，コードはネットワークの構造定義とパラメーターを設定するものであった．コードの最後の行は実際

3) フラットベクトルは，多次元配列を1次元ベクトルに変換したベクトルである．

network. The last lines of code actually create the neural network and run it with the "Adadelta" optimizer:

```
model.compile(loss=keras.losses.categorical_crossentropy,
optimizer=keras.optimizers.Adadelta(), metrics=['accuracy'])
model.fit(Xtrain, Otrain, batch_size=batch_size, epochs=epochs,
verbose=1, validation_data=(Xtest, Otest))
score = model.evaluate(Xtest, Otest, verbose=0)
print('Test loss:', score[0])
print('Test accuracy:', score[1])
```

The procedure model.fit trains the model and finds the best fitting parameters. The network has now *learned* how to recognize the 10 hand-written characters from the 60000 examples we provided. The function model.evaluate uses 10000 new, previously unseen examples to test the model and evaluates how good it is with unknown data.

With few comments and minor aesthetic modifications, the complete code is summarized here:

Listing 5.1 Example of a CNN that learns to recognize hand-written digits. Written in python with the Keras command library. To understand every details of this program you need to study the Keras manual.

```
import keras
from keras.datasets import mnist
from keras.models import Sequential
from keras.layers import Dense, Dropout, Flatten
from keras.layers import Conv2D, MaxPooling2D
from keras.utils.np_utils import to_categorical
from numpy import *
# read the MNIST dataset
(Xtrain, Otrain), (Xtest, Otest) = mnist.load_data()
print shape(Xtrain),shape(Xtest)
print shape(Otrain),shape(Otest)
# convert to 2D "pictures"
Xtrain = Xtrain.reshape(60000, 28,28,1)
Xtest = Xtest.reshape(10000, 28,28,1)
# convert to correct type
```

にニューラルネットワークを作成し，“Adadelta” という最適化の手法で実行する．

```
model.compile(loss=keras.losses.categorical_crossentropy,
optimizer=keras.optimizers.Adadelta(), metrics=['accuracy'])
model.fit(Xtrain, Otrain, batch_size=batch_size, epochs=epochs,
verbose=1, validation_data=(Xtest, Otest))
score = model.evaluate(Xtest, Otest, verbose=0)
print('Test loss:', score[0])
print('Test accuracy:', score[1])
```

model.fit 関数によって最適なパラメーターを探し出す．ネットワークは6万の教師データから10種類の手書きの画像を学習した．そして，model.evaluate 関数により，1万の未知のデータを判定し，同モデルの未知のデータへのはたらきを検証する．ここまでの説明にいくつかの付け足しと修正を加えてコードを完成させる必要がある．

リスト 5.1　CNN を用いた手書き数字の識別学習例．Python の Keras ライブラリーを使用した．プログラムの詳細の理解には，Keras のマニュアルを参照することをお薦めする．

```
import keras
from keras.datasets import mnist
from keras.models import Sequential
from keras.layers import Dense, Dropout, Flatten
from keras.layers import Conv2D, MaxPooling2D
from keras.utils.np_utils import to_categorical
from numpy import *
# read the MNIST dataset
(Xtrain, Otrain), (Xtest, Otest) = mnist.load_data()
print shape(Xtrain),shape(Xtest)
print shape(Otrain),shape(Otest)
# convert to 2D "pictures"
Xtrain = Xtrain.reshape(60000, 28,28,1)
Xtest = Xtest.reshape(10000, 28,28,1)
# convert to correct type
```

```
Xtrain = Xtrain.astype('float32') / 255
Xtest = Xtest.astype('float32') / 255
# convert to categories
Otrain = to_categorical(Otrain)
Otest = to_categorical(Otest)
print shape(Otrain),shape(Otest)
print shape(Xtrain),shape(Xtest)
print(Xtrain.shape[0], 'train samples')
print(Xtest.shape[0], 'test samples')
batch_size = 128
num_classes = 10
epochs = 1
model = Sequential()
model.add(Conv2D(32, kernel_size=(3, 3),
activation='relu', input_shape=(28,28,1)))
model.add(Conv2D(64, (3, 3), activation='relu'))
model.add(MaxPooling2D(pool_size=(2, 2)))
#model.add(Dropout(0.25))
model.add(Flatten())
model.add(Dense(128, activation='relu'))
#model.add(Dropout(0.5))
model.add(Dense(num_classes, activation='softmax'))
model.compile(loss=keras.losses.categorical_crossentropy,
optimizer=keras.optimizers.Adadelta(), metrics=['accuracy'])
model.fit(Xtrain, Otrain, batch_size=batch_size, epochs=epochs,
verbose=1, validation_data=(Xtest, Otest))
score = model.evaluate(Xtest, Otest, verbose=0)
print('Test loss:', score[0])
print('Test accuracy:', score[1])
```

If you run this on your PC, the neural network should reach the ability to recognize hand written characters in few minutes! The printout of the "test accuracy" should result to be better than 97%. The model now is trained and can be re-used for ever to recognize characters when you need it. You can always save your trained networks, with the Keras command model.save().

```
Xtrain = Xtrain.astype('float32') / 255
Xtest = Xtest.astype('float32') / 255
# convert to categories
Otrain = to_categorical(Otrain)
Otest = to_categorical(Otest)
print shape(Otrain),shape(Otest)
print shape(Xtrain),shape(Xtest)
print(Xtrain.shape[0], 'train samples')
print(Xtest.shape[0], 'test samples')
batch_size = 128
num_classes = 10
epochs = 1
model = Sequential()
model.add(Conv2D(32, kernel_size=(3, 3),
activation='relu', input_shape=(28,28,1)))
model.add(Conv2D(64, (3, 3), activation='relu'))
model.add(MaxPooling2D(pool_size=(2, 2)))
#model.add(Dropout(0.25))
model.add(Flatten())
model.add(Dense(128, activation='relu'))
#model.add(Dropout(0.5))
model.add(Dense(num_classes, activation='softmax'))
model.compile(loss=keras.losses.categorical_crossentropy,
optimizer=keras.optimizers.Adadelta(), metrics=['accuracy'])
model.fit(Xtrain, Otrain, batch_size=batch_size, epochs=epochs,
verbose=1, validation_data=(Xtest, Otest))
score = model.evaluate(Xtest, Otest, verbose=0)
print('Test loss:', score[0])
print('Test accuracy:', score[1])
```

紹介したコードを実装すれば，短時間で手書き数字の学習や識別を体験することができる．“test accuracy” は精度の出力で，正答率が 97% 以上になる．また，これらのコードによりいつでも繰り返し分類が可能になる．model.save() によって，いつでも学習済みのネットワークを実装することが可能である．もちろん正答率は保証されている．

Great result indeed.

Think about these questions

- What is the mathematical expression of a convolution?
- What is a *filter* and how is it mathematically expressed?
- Can you name some example of filters?
- What is the *pooling* process?
- Is pooling also a convolution or not?
- Can optimizer also alter filters or not?
- Why are CNNs better than conventional ANNs?
- Make a program that realizes the same network presented in chapter 4, by using the library Keras.

考えてみよう

- 畳み込み関数を数式で表すとどうなるか？
- フィルターを数式で表すとどうなるか？
- フィルターの例はどのようなものか？
- プーリング処理とはどのようなものか？
- プーリング処理も畳み込み処理の一種か？
- 最適化関数によりフィルターの値を変更することはできるか？
- なぜ CNN は，従来の ANN に比べて優れているのか？
- 第 4 章で構成したネットワークを Keras でも作ってみよう．

Chapter 6

Recurrent neural networks

Recurrent neural networks (RNN) are networks that somehow connect their output to their input. They contain a kind of information feedback. Their output is not only function of the current input, but also of previous outputs. In a sense, these networks have *memory*. To understand RNN in more depth, we should have a look at one of the first recurrent networks: the Hopfield network.

6.1 The Hopfield network

Imagine you have a neuron layer in which each neuron is connected to all the other neurons. To be more clear, if we have N neurons every of them has an input and an output: the input receives the outputs of all the other $N-1$ neurons and the output is connected to all the $N-1$ inputs. This fully connected system is called a *recurrent* layer (see an example in figure 6.1).

Using this structure, Hopfield[Hopfield, 1982] constructed a network able to memorize patterns. Let's see how in details. Suppose that each neuron is *bi-stable*, it has only two output states. The output of neuron X can be only 1 or -1. If we have N neurons i, we can write $X_i \in \{-1, 1\}^N$. Also we will have, for each neuron X_i, $(N-1)$ connections $W_{ij}^{N^2} \in \{-1, 1\}$. As in a biological neuron, every neuron output should be proportional to the sum of all its inputs. In the Hopfield network we have only two possible output values, 1 or -1, so we do not need a squeezing function like in chapter 4. Very simply, each neuron X_i is subject to this equation

第6章

フィードバックをもつニューラルネットワーク

この章で取り扱う再帰型ニューラルネットワーク (recurrent neural networks: RNN) は出力をいかに入力に戻すかということが重要なネットワークである．つまり，情報のフィードバックをもつネットワークと言うことができる．出力は単に現状の入力の関数というわけでなく，過去の出力もまた入力に含まれる．つまり，記憶をもつネットワークということもできるだろう．RNN をより理解するために，まず最初の再帰型のネットワークとして知られているホップフィールドネットワークについて調べてみよう．

6.1 ホップフィールドネットワーク

まず，各々のニューロンが他のすべてのニューロンと結合しているようなニューロンの層をイメージしてみよう．正確に言えば，入力と出力を含む N 個のニューロンの層である．入力は他の $N-1$ 個の出力と結合し，出力も他の $N-1$ 個の入力と結合する．この全相互結合型のシステムを，再帰型の層と呼ぶ（図 6.1)．

ホップフィールドはこのネットワーク構造を用いることで，パターン記憶可能なネットワークを構築した [Hopfield, 1982]．次に詳細を示す．

まずすべてのニューロンが 2 状態モデル[1]と仮定すると，出力も 2 つの状態のみもつことになる．出力ニューロン X は，1 か -1 の値のみ許されている．N 個のニューロンを i で識別することで，$X_i \in \{-1,1\}^N$ と記述することができる．また，各ニューロン X_i に対して $N-1$ 個の結合荷重 $W_{ij} \in \{-1,1\}^{N^2}$ を設定する．実際の生物学的なニューロンの場合でも，すべてのニューロンの出力が入力の合計値に比例しているべきだと言うことができる．

1) 2 状態モデル：bi-stable, two state model, 2 状態モデルは，統計力学の分野で広く多用される．

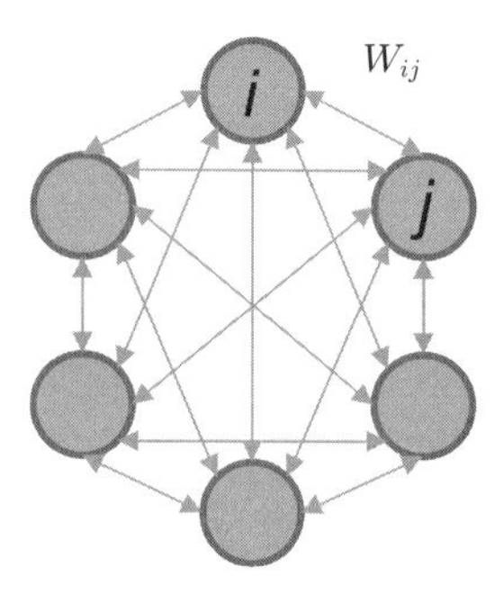

Fig. 6.1 A sketch of a Hopfield network composed by $N = 6$ neurons only.

$$\begin{aligned} H_i &= \Sigma_{j=1}^{N} W_{ij} X_j + b_i \\ \text{if } H_i > 0 &: X_i = 1 \text{ else} : X_i = -1 \end{aligned} \tag{6.1}$$

where b_i is the bias and H_i is the total input for neuron i (this is also called the *field at* i). For simplicity let's choose again $b_i = 0$. Since X_i can assume only two values, we impose that if the value H_i is bigger than zero, $X_i = 1$ otherwise $X = -1$. Equation 6.1 implies that given a random initial condition $X_i^0 \in \{-1, 1\}$ and a set of random weights W_{ij}, the system will be unstable. The network will move away from its initial pattern $\bar{X}_0$ to a different one $\bar{X}_1$. The state $\bar{X}_1$ will be the new input and this, using the same weights W_{ij}, will drive the network to the next state $\bar{X}_2$ and so on. In general, all these states will be different to each other. We say that the network is *unstable.*

Is there a way to choose W_{ij} in order to make the network stable? Hopfield found, and you can verify very easily, that for a given pattern $X_i^m \in \{-1, 1\}^N$, the choice of

$$W_{ij} = X_i^m X_j^m \tag{6.2}$$

will yield a stable pattern. In this way if the two values X_i and X_j are the same, their product W_{ij} will be 1 (it will be -1 if they are different). This will result in a stable output equal to the pattern X_i^m. In fact

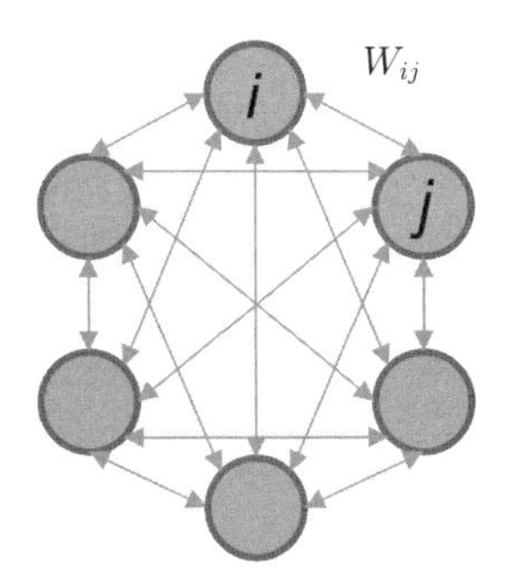

図 6.1　$N = 6$ のニューロンによるホップフィールドネットワーク概略図.

ホップフィールドネットワークでは，1 か -1 の 2 つの状態しか存在しないため，第 4 章のような活性化関数は必要ない．簡単に言うと各ニューロン X_i は下記の式で表される．

$$\begin{gathered} H_i = \Sigma_{j=1}^{N} W_{ij} X_j + b_i \\ \text{if } H_i > 0 : X_i = 1 \text{ else} : X_i = -1 \end{gathered} \tag{6.1}$$

ここで b_i はバイアスである．また，H_i は各ニューロンの入力の合計値である（i の場とも呼ばれる）．まず，単純化のためにここでもバイアスは $b_i = 0$ とする．X_i は 2 状態を仮定しているので H_i の値が 0 より大きいとき，$X_i = 1$ となる．また，それ以外のとき，$X_i = -1$ となる．式 6.1 は初期状態 $X_i^0 \in \{-1, 1\}$ とランダムに値を設定した結合荷重 W_ij を与えた場合システムはとても不安定[2)]な状態になることを示唆する．そして，すぐに最初のパターン $\bar{X}_0$ から異なるパターン $\bar{X}_1$ へと遷移するだろう．新たな状態 $\bar{X}_1$ と結合荷重 W_ij により次の状態 $\bar{X}_2$ へと遷移しその後同じように続く．一般的に言えばすべての状態は互いに異なっている．このようなネットワークを不安定なネットワークと呼ぶ．

ネットワークを安定化するために適切な結合荷重 W_ij は存在するのだろうか？　ホップフィールドはパターン $X_i^m \in \{-1, 1\}$ を与えるという，とても簡単な次の方法でこのことを検証した．

2) 不安定:unstable，ネットワークの出力が収束していない状態．

$$H_i = \Sigma_{j=1}^{N} W_{ij} X_j^m = \Sigma_{j=1}^{N} X_i^m X_j^m X_j^m \tag{6.3}$$

consider that X is limited only to 1 or -1. This implies that $X_j^m\ X_j^m$ is always one, the field in X_i is

$$H_i = \Sigma_{j=1}^{N} X_i^m = (N-1) X_i^m \tag{6.4}$$

Because of the rules in equation 6.1, the new X_i^m will be equal to the old X_i^m. The network state does not change, it is stable. We can say that the *pattern* $\bar{X}^m = \{X_i^m\}$ is a stable *point* of the network. The condition given by equation 6.2 is also called *Hebbian rule* because it is inspired by the behavior of real biological neurons.[1)]

Now, suppose we have two or more patterns (you can think that each neuron is a pixel and the pattern an *image*) $\bar{X}^p$ with $p \in \{1, 2, 3 ... N_l\}$, where N_l is the number of patterns. We want to store these images in the memory of our network. Can we choose the weights W_{ij} appropriately in order to memorize these pattern? Surely, we can! We choose the weights with a *generalized* form of the Hebbian rule:

$$W_{ij} = \frac{1}{N} \Sigma_p X_i^p X_j^p \tag{6.5}$$

You can verify yourself that these weights force the network to converge to stable conditions. In fact, if we calculate the field for a pattern m

$$H_i^m = \Sigma_{j=1}^{N} \left(\frac{1}{N} \Sigma_p X_i^p X_j^p\right) X_j^m \tag{6.6}$$

we notice that if m is one of those p patterns memorized, then $X_j^p X_j^m = 1$. Thus, we can split equation 6.6 in

1) In biological neurons, the connection strength gets stronger when the two neurons are spiking in synchronous, whereas it gets weaker if they don't. Here equation 6.2 says a similar thing: when the outputs of two neurons are multiplied, multiplication in mathematics can be thought as synchronicity, the two neuron connection strength become maximum.

$$W_{ij} = X_i^m X_j^m \tag{6.2}$$

もし X_i と X_j が同じであるなら，それらの積 W_{ij} は 1 となり，異なる場合は –1 となる．この式による W_ij を用いることでパターン X_i^m と等しい安定した出力を得ることができる．

$$H_i = \Sigma_{j=1}^N W_{ij} X_j^m = \Sigma_{j=1}^N X_i^m X_j^m X_j^m \tag{6.3}$$

実際，X の値は –1 か 1 に制限されていると仮定しているので，X_j^m X_j^m の値は 2 乗されて常に 1 になる．X_i^m は次式で表される．

$$H_i = \Sigma_{j=1}^N X_i^m = (N-1) X_i^m \tag{6.4}$$

上式は式 6.1 より導かれる．更新された X_i^m は前の状態の X_i^m と等しくなる．つまり，そのネットワークの状態は変わらず安定しているということになる．パターン $\bar{X}^m = \{X_i^m\}$ がネットワーク安定化の条件と言うことができる．式 6.2 で表される条件はヘッブの法則と呼ばれている．この法則は実際の生物学的なニューロンのふるまいにヒントを得て立てた仮説である[3)]．今ここに $\bar{X}^p$ $p \in \{1, 2, 3 ... N_l\}$ で表される 2 つ以上の画像パターンがある（画像における 1 ピクセルが 1 ニューロン，パターンが画像とする）．N_l は，画像パターンの数であるとする．これらの画像をネットワークのメモリーに保存することを考える．画像パターンを記憶するために適切な結合荷重 W_{ij} を選ぶことはできるのだろうか？　もちろん可能である．ヘッブの法則を用いて結合荷重を最適化する．

$$W_{ij} = \frac{1}{N} \Sigma_p X_i^p X_j^p \tag{6.5}$$

ヘッブの法則を用いて結合荷重を最適化することで，ネットワークが安定状態へ収束することを検証してみよう．実際に画像 m を計算してみよう．

3) 生物学的なニューロンは，同時に発火するとき，その結合荷重はより強くなる．また，同時発火がなければ弱くなる．このことは，式 6.2 に表されている．つまり，2 つのニューロンの出力が掛け合わされたとき，数学的には積の作用は同期として考えることができるため，それらのニューロンの結合荷重は最大になる．

$$H_i^m = \frac{1}{N}\Sigma_{j=1}^N (X_i^m + \Sigma_{p\neq m} X_i^p X_j^p X_j^m)$$
$$H_i^m = X_i^m + \frac{1}{N}\Sigma_{j=1}^N \Sigma_{p\neq m} X_i^p X_j^p X_j^m \tag{6.7}$$

The term $\frac{1}{N}\Sigma_{j=1}^N \Sigma_{p\neq l} X_i^p X_j^p X_j^m$ is called *crosstalk* and if it is smaller than one, H_i^m becomes equal to the pattern X_i^m. This is valid for any m. So if the crosstalk is not too big, the network have memorized all our patterns. We now understand that we cannot store too many patterns in the network, otherwise the crosstalk will get too big. To increase the store *capacity* of our network, we must increase the total number of neurons N. It can be demonstrated that the ratio between the total number of patterns memorized N_l and the network neurons N must be smaller than about 0.14 to avoid crosstalk, $N_l/N \leq 0.14$ (see for example prof. A. Edalat[Edalat, 2019]).

Let's try to make a simple Hopfield network right here, so you can play with it by yourself. We use python language again. First we import the numerical libraries for matrix multiplication and the graphics libraries for easy plotting

```
from numpy import *
import matplotlib.pyplot as plt
```

then we make a procedure to plot an image:

```
def plotImg(img,tt):
        # show the image
        plt.imshow(img)
        plt.title(tt)
        plt.colorbar()
        plt.show()
```

We will use this to visualize our patterns. The arguments *img* is the pattern to visualize and *tt* is a convenient title that we can give to the graph. Now

$$H_i^m = \Sigma_{j=1}^N (\frac{1}{N}\Sigma_p X_i^p X_j^p) X_j^m \tag{6.6}$$

記憶された p 個のパターンの中の 1 つは m であった場合，$X_j^p X_j^m = 1$ である．つまり，式 6.6 を場合分けすると次式のように表される．

$$\begin{aligned} H_i^m &= \frac{1}{N}\Sigma_{j=1}^N (X_i^m + \Sigma_{p\neq m} X_i^p X_j^p X_j^m) \\ H_i^m &= X_i^m + \frac{1}{N}\Sigma_{j=1}^N \Sigma_{p\neq m} X_i^p X_j^p X_j^m \end{aligned} \tag{6.7}$$

$\frac{1}{N}\Sigma_{j=1}^N \Sigma_{p\neq l} X_i^p X_j^p X_j^m$ の項は相互干渉と呼ばれ，その値が 1 より小さい場合，H_i^m は X_i^m と等しくなる．ここで，どの画像パターン m に対しても有効なため，相互干渉の値が小さい場合，ネットワークはすべての入力パターンを記憶する．現状ではパターンがあまりにも多い場合，相互干渉の値が大きくなり記憶することができない．ネットワークの記憶容量を増やすためには，扱うニューロンの数 N を増やさなければならない．相互干渉を避けるためには，記憶されたパターンの数 N_l とニューロンの数 N の比が 0.14 以下でなくてはならない．$N_l/N \leq 0.14$（[Edalat, 2019] を参照）．

再び Python を使って実際にホップフィールドネットワークを実装してみよう．まずは，行列計算や可視化のために，数値解析ライブラリーと描画ライブラリーをインポートしよう．

```
from numpy import *
import matplotlib.pyplot as plt
```

次に画像データを描画する関数を設定する．

```
def plotImg(img,tt):
        # show the image
        plt.imshow(img)
        plt.title(tt)
        plt.colorbar()
        plt.show()
```

この関数によって画像データを可視化する．ここで引数 img は可視化させるための画像データで引数 tt はグラフのタイトルである．ネットワークに記

we define two images to be stored in the network. We make a very simple square and a plus in a 5×5 pixel matrix:

```
# a "rectangle" to store
rX=[
[1,1,1,1,1],
[1,0,0,0,1],
[1,0,0,0,1],
[1,0,0,0,1],
[1,1,1,1,1],
]
rX=array(rX) # transform to array
rX[rX<=0]=-1 # put all zeros to -1
# a "cross" to store
cX=[
[0,0,1,0,0],
[0,0,1,0,0 ],
[1,1,1,1,1],
[0,0,1,0,0],
[0,0,1,0,0],
]
cX=array(cX) # transform to array
cX[cX<=0]=-1 # put all zeros to -1
```

For visual clarity, we constructed the image vector with ones and zeros, then the zeros are converted to -1 using *logical indexing* in the array assignments $rX[rX <= 0] = -1$, and similarly for the other image cX. Now that we have two images, let's give the network an initial random pattern:

```
X=random.randint(2,size=(5,5))
X[X<=0]=-1 # put all zeros to -1
plotImg(X,"the initial image")
```

The first two lines create a random 5×5 arrays of ones and zeros, convert all zeros to -1s then our procedure visualizes it. You will see a window appear with a random pattern of bright and dark dots. Now we will create array W_{ij} in the same way and visualize it.

憶させる2つの画像を定義する．とても簡単な5×5の正方行列を用意する(四角と十字)．

```
# a "rectangle" to store
rX=[
[1,1,1,1,1],
[1,0,0,0,1],
[1,0,0,0,1],
[1,0,0,0,1],
[1,1,1,1,1],
]
rX=array(rX) # transform to array
rX[rX<=0]=-1 # put all zeros to -1
# a "cross" to store
cX=[
[0,0,1,0,0],
[0,0,1,0,0 ],
[1,1,1,1,1],
[0,0,1,0,0],
[0,0,1,0,0],
]
cX=array(cX) # transform to array
cX[cX<=0]=-1 # put all zeros to -1
```

簡潔な表現のために，画像データは0と1のベクトルで表現する．さらに，配列の引数として論理式[4] $rX[rX <= 0] = -1$ を用いることで0を-1に変換する．同様な処理をもう1つの画像データcXにも行う．ここまでで2つの画像データが準備できた．次にネットワークの初期値を乱数を用いて設定してみよう．

```
X=random.randint(2,size=(5,5))
X[X<=0]=-1 # put all zeros to -1
plotImg(X,"the initial image")
```

まず，最初の2行により，1の組み合わせで構成される5×5の配列が作成される．その後再び同じ関数を用いて，0を-1に変換する．ランダムに配置

4) 論理式：logical indexing，ホップフィールドの仮定から入力画像は-1か1の組み合わせの必要がある．

```
W=random.randint(2,size=(25,25))
W=array(W) # transform to array
W[W<=0]=-1 # put all zeros to -1
fill_diagonal(W,0) # Wij is zero if i=j !
plotImg(W,"the random weights")
```

Note that neurons are not connected to themselves, so W_{ij} must be zero if $i = j$. The numpy library command "fill_diagonal(W,0)" does this for us. Using this random weights $\bar{W}$ we now apply the Hopfield rules to the network. We calculate the field and update all the neuron outputs $\bar{X}$.

```
# apply the Hopfield rule
X=X.flatten() # transform to vector
H=dot(W,X) # field (5*5=25 vector)
X[H>0] = 1    # Hopfield rule
X[H<=0]=-1    # Hopfield rule
plotImg(X.reshape(5,5),"result image")
```

We cannot multiply two matrix of different sizes, so the dot product is between a 25×25 array (W) and the 25 elements long one-dimensional vector representing our 5×5 pixels pattern X. The conversion is done by the "X.flatten()" command. The plot will result in another random pattern of pixels. But what happens if we store some information in the network? Let's choose the two simple 5×5 pixels images:

```
plotImg(rX,"an image to learn")
plotImg(cX,"an image to learn")
```

By plotting these two lines above, you can recognize a square and a sort of plus sign (a cross). These images will be stored in the connection weights W using the generalized Hebbian rule in 6.5

```
rX=rX.flatten();cX=cX.flatten() # the two images
for i in range(25):
```

された白黒の画像データを確認することができる．結合荷重 W_{ij} についても同様に配列を作り可視化してみよう．

```
W=random.randint(2,size=(25,25))
W=array(W) # transform to array
W[W<=0]=-1 # put all zeros to -1
fill_diagonal(W,0) # Wij is zero if i=j !
plotImg(W,"the randomweights")
```

各々のニューロンは自分自身と結合することができないことに注意する．もし，$i=j$ なら，Wij は 0 でなければならない．Numpy ライブラリーのコマンド "fill_diagonal(W,0)" はそれらの操作を行う．ランダムな結合荷重 W を用いることで，ホップフィールドの規則をネットワークに適用させる．ネットワークの計算を終え，すべての更新が済むとニューロンは $\bar{X}$ を出力する．

```
# apply the Hopfield rule
X=X.flatten() # transform to vector
H=dot(W,X) # field (5*5=25 vector)
X[H>0] = 1   # Hopfield rule
X[H<=0]=-1   # Hopfield rule
plotImg(X.reshape(5,5),"result image")
```

異なるサイズの行列同士を掛け合わせることはできないため，25×25 の配列 $\bar{W}$ と 25 の要素をもつ 1 次元ベクトルの積を計算する．ここでは，5×5 のパターン $\bar{X}$ を 25 の要素をもつ 1 次元ベクトルに変換している．変換は，コマンド "X.flatten()" によって行われる．プロットの結果は，また別のランダムなパターンになるかもしれない．ネットワークにいくつかの情報を記憶させると何が起こるのだろうか？ 2 つの簡単な画像データを記憶させてみよう．

```
plotImg(rX,"an image to learn")
plotImg(cX,"an image to learn")
```

上記の 2 つのデータをプロットすると四角と十字のデータであることがわかる．それらの画像は，式 6.5 のヘッブの法則によって結合荷重 W を通じてネットワークに記憶される．

```
rX=rX.flatten();cX=cX.flatten() # the two images
for i in range(25):
```

```
        for j in range(25):
                #  Wij=1/2 (xi*xj+yi*yj)
                W[i,j]=(0.5)*(rX[i]*rX[j]+cX[i]*cX[j])
plotImg(W,"the connections W")
```

Notice that we had to convert our 5×5 images in one dimensional vector again. The last line plots the new array W. Noticeably it has a geometrical structure that represents the information contained in the two images. Now let's test the network. Has it *memorized* the two patterns? Let's start again from a random initial condition $\bar{X}$

```
# start with a random pattern
X=random.randint(2,size=(5,5))
X[X<=0]=-1 # put all zeros to -1
plotImg(X,"the start image X")
```

then we apply the Hopfield rule again

```
X=X.flatten() # transform to vector
H=dot(W,X) # field (5*5=25 vector)
X[H>0] = 1    # Hopfield rule
X[H<=0]=-1    # Hopfield rule
plotImg(X.reshape(5,5),"the netOutput")
```

You will see that the network will output to one of the two images, the Hopfield network has *memory*! Depending on the input pattern $\bar{X}$, the network converges to one of the memorized images, the one that is more similar to the input. This is called *auto-associative* propriety and it is a characteristics of human memory too.

Notice that the information is stored in the $\bar{W}$, the connections strength between neurons. This fact is reflected in real biological brains: the information, all that the human brain knows and all the brain functions are implicit in the inter-neuronal *synaptic strength*. All the learning and the dynamism of information flow occurs because of changes in synaptic strength. Hopfield model is very simple, nevertheless it is used as a

```
        for j in range(25):
                #  Wij=1/2 (xi*xj+yi*yj)
                W[i,j]=(0.5)*(rX[i]*rX[j]+cX[i]*cX[j])
plotImg(W,"the connections W")
```

再び，5×5 の画像データを 1 次元のベクトルに変換しなければならないことに注意しよう．最終行では，更新後の配列 W をプロットしている．そこには，2 つの入力画像データの情報を含んだ幾何学的な構造がはっきりと見えてくる．ネットワークを実装してみよう！　2 つのパターンを記憶することができただろうか？　パターン $\bar{X}$ を初期化してもう一度ネットワークを実装してみよう！

```
# start with a random pattern
X=random.randint(2,size=(5,5))
X[X<=0]=-1 # put all zeros to -1
plotImg(X,"the start image X")
```

次に，再度ホップフィールドの法則を適応して，次のようになる．

```
X=X.flatten() # transform to vector
H=dot(W,X) # field (5*5=25 vector)
X[H>0] = 1    # Hopfield rule
X[H<=0]=-1    # Hopfield rule
plotImg(X.reshape(5,5),"the netOutput")
```

ネットワークが 2 つの画像のうち，どちらか 1 つの画像を出力することが確認できる．つまりホップフィールドネットワークは記憶からの出力を実現した．入力パターン $\bar{X}$ によってネットワークは，記憶された画像の中で，入力により近いどれか 1 つに収束する．これは連想記憶型[5)]と呼ばれ，人間の記憶の仕組みの特徴に類似している．

情報はニューロン同士の結合荷重 $\bar{W}$ に蓄積されていることに注目してみよう．このことは，人間の脳にも同じような現象が確認できる．脳に情報が伝達されると，脳の内部でシナプス結合として潜在的に機能する．すべての学習と情報の動的な流入は，シナプスの結合力の変化を引き起こす．ホップフィール

5) 連想記憶型：auto-associative propriety，学習を通じて自動で連想が行われるネットワークが形成される．

memory paradigm for behavioral/cognitive prototypes in psychology and psychotherapy[Edalat, 2019].

Here is the complete listing of our 5×5 neurons Hopfield network, with few comments and aesthetic modifications:

Listing 6.1 A test of a Hopfield network that learns two simple patterns in the language Python.

```
from numpy import *
import matplotlib.pyplot as plt
def plotImg(img,tt):
        # show the image
        plt.imshow(img)
        plt.title(tt)
        plt.colorbar()
        plt.show()
# two "images" to store
rX=[
[1,1,1,1,1],
[1,0,0,0,1],
[1,0,0,0,1],
[1,0,0,0,1],
[1,1,1,1,1],
]
rX=array(rX) # transform to array
rX[rX<=0]=-1 # put all zeros to -1
cX=[
[0,0,1,0,0],
[0,0,1,0,0],
[1,1,1,1,1],
[0,0,1,0,0],
[0,0,1,0,0],
]
cX=array(cX) # transform to array
cX[cX<=0]=-1 # put all zeros to -1
# a random image
X=random.randint(2,size=(5,5))
X[X<=0]=-1 # put all zeros to -1
```

ドモデルはとてもシンプルだが，心理学や精神療法における行動や認知にとっての記憶の仕組みとして有効なモデルである [Edalat, 2019]．ここに，5×5 のニューロンを使ったホップフィールドネットワークのアルゴリズムを，コメントと共に示す．

リスト 6.1 Python 言語を用いた２つの簡単なパターンのホップフィールドネットワーク学習例

```
from numpy import *
import matplotlib.pyplot as plt
def plotImg(img,tt):
        # show the image
        plt.imshow(img)
        plt.title(tt)
        plt.colorbar()
        plt.show()
# two "images" to store
rX=[
[1,1,1,1,1],
[1,0,0,0,1],
[1,0,0,0,1],
[1,0,0,0,1],
[1,1,1,1,1],
]
rX=array(rX) # transform to array
rX[rX<=0]=-1 # put all zeros to -1
cX=[
[0,0,1,0,0],
[0,0,1,0,0],
[1,1,1,1,1],
[0,0,1,0,0],
[0,0,1,0,0],
]
cX=array(cX) # transform to array
cX[cX<=0]=-1 # put all zeros to -1
# a random image
X=random.randint(2,size=(5,5))
X[X<=0]=-1 # put all zeros to -1
```

```
plotImg(X,"the initial image")
# a set of random weights
W=random.randint(2,size=(25,25))
W=array(W) # transform to array
W[W<=0]=-1 # put all zeros to -1
fill_diagonal(W,0) # Wij is zero if i=j !
plotImg(W,"the randomweights")
# apply the Hopfield rule
X=X.flatten() # transform to vector
H=dot(W,X) # field (5*5=25 vector)
X[H>0] = 1   # Hopfield rule
X[H<=0]=-1   # Hopfield rule
plotImg(X.reshape(5,5),"result image")
# now store two images
plotImg(rX,"an image to learn")
plotImg(cX,"an image to learn")
# make the Wij=1/2 (xi*xj+yi*yj)
rX=rX.flatten();cX=cX.flatten() # the two images
for i in range(25):
        for j in range(25):
                #  Wij=1/2 (xi*xj+yi*yj)
                W[i,j]=(0.5)*(rX[i]*rX[j]+cX[i]*cX[j])
plotImg(W,"the connections W")
# start with a random pattern
X=random.randint(2,size=(5,5))
X[X<=0]=-1 # put all zeros to -1
plotImg(X,"the start image X")
X=X.flatten() # transform to vector
H=dot(W,X) # field (5*5=25 vector)
X[H>0] = 1   # Hopfield rule
X[H<=0]=-1   # Hopfield rule
plotImg(X.reshape(5,5),"the netOutput")
```

In this Hopfield network, we calculated the weights $\bar{W}$s *a priori* with the Hebbian rule 6.5. So, given an input $\bar{X}_0$, a single calculation produces the output $\bar{Y}_0$ for the complete network. This output will be the next input! So it seems that the Hopfield network is a sort of *synchronous* machine:

```
plotImg(X,"the initial image")
# a set of random weights
W=random.randint(2,size=(25,25))
W=array(W) # transform to array
W[W<=0]=-1 # put all zeros to -1
fill_diagonal(W,0) # Wij is zero if i=j !
plotImg(W,"the randomweights")
# apply the Hopfield rule
X=X.flatten() # transform to vector
H=dot(W,X) # field (5*5=25 vector)
X[H>0] = 1   # Hopfield rule
X[H<=0]=-1   # Hopfield rule
plotImg(X.reshape(5,5),"result image")
# now store two images
plotImg(rX,"an image to learn")
plotImg(cX,"an image to learn")
# make the Wij=1/2 (xi*xj+yi*yj)
rX=rX.flatten();cX=cX.flatten() # the two images
for i in range(25):
        for j in range(25):
                #  Wij=1/2 (xi*xj+yi*yj)
                W[i,j]=(0.5)*(rX[i]*rX[j]+cX[i]*cX[j])
plotImg(W,"the connections W")
# start with a random pattern
X=random.randint(2,size=(5,5))
X[X<=0]=-1 # put all zeros to -1
plotImg(X,"the start image X")
X=X.flatten() # transform to vector
H=dot(W,X) # field (5*5=25 vector)
X[H>0] = 1   # Hopfield rule
X[H<=0]=-1   # Hopfield rule
plotImg(X.reshape(5,5),"the netOutput")
```

ホップフィールドネットワークではそれぞれの結合荷重 $\bar{W}$ を演繹的に[6]ヘッブの法則によって計算する．つまり，入力 $\bar{X}_0$ を与えると，計算によって出力 $\bar{Y}_0$ を得る．この出力が次の入力になる．このことからホップフィールドネ

6) 演繹的に：priori.

it calculates an output $\bar{Y}_0$ in one time-step, then this becomes the next time-step input $\bar{X}_0$.

You understand that this cannot be what is happening in biological brains. Real brains are *asynchronous* machines: calculations are done at random time-steps. This means that every neuron updates its values in random moments in time. It is possible to simulate that in a Hopfield network very easily. Just apply equation 6.5 at random moments in time. You can have fun and try to modify listing 6.1 to achieve asynchronous updating like in a real brain: the network will work fine!

6.2 Complex recurrent neural networks

In the Hopfield network each neuron output becomes the input of all the other neurons. This is a sort of *recursive* situation, the word *recurrent* refers to this *feedback* structure. Because the Hopfield network is fully recursive and has the ability to memorize patterns, we can infer that recursion has something to do with memory. This fact has been studied by many researchers and mathematicians. They put forward a more efficient layered recursive structure, like the one in figure 6.2.

In this case a layer of neurons works like a standard ANN of chapter 4, but its output is fed back to the input. In other words, the input of the network is not only the usual vector $\bar{X}_0$, but also the output $\bar{Y}_0$. I presume you are feeling there is a problem here: what about timing? If the input contains the output, how can we calculate the output if we do not yet have the whole input?

This is exactly the same problem as with the Hopfield network, but it happens to the layers instead that to the single neurons. The solution, again, is to update the values in a *synchronous* way. That is, at one time step the output $\bar{Y}_0$ is calculated from $\bar{X}_0$ and nothing else. Then at the next time step, the network will have two new inputs: $\bar{X}_1$ and $\bar{Y}_0$ and this will produce $\bar{Y}_1$. This will be the next input together with $\bar{X}_2$ and so on.

ットワークは同期型の機械の一種と言われる．つまり，あるタイムステップでの出力 $\bar{Y}_0$ は，次のタイムステップの入力 $\bar{X}_0$ となる．

しかし，現実の脳ではこのような状況が起こることはない．実際の脳は非同期型だからである．つまり，計算はランダムなタイムステップで処理される．これは，すべてのニューロンがランダムなタイミングで更新されることを意味する．ホップフィールドネットワークでは，とても簡単にシミュレート可能である．それは，単に式 6.5 の時間に関してランダムに更新すれば良いのである．リスト 6.1 を本物の脳のように非同期に更新するコードに修正してみよう．非同期に修正してもネットワークはきちんと動くだろう．

6.2　再帰型ニューラルネットワーク

ホップフィールドネットワークでは，各ニューロンの出力は，他のすべてのニューロンの入力となる．これは再帰型の構造に分類される．再帰型[7)]構造とは，フィードバック構造を示す．ホップフィールドネットワークは，すべての信号を再帰的に扱い，パターンを記憶するため，再帰型は記憶を扱うことに関係があると推測できる．実際，多くの研究者や数学者がこのことについて研究している．その結果，再帰型多層構造がより有力である．図 6.2 にその概略を示した．

このケースでは，第 4 章の標準的な ANN のように一層のニューロンがはたらくが，出力が入力のフィードバックになる．言い換えれば，ネットワークの入力は $\bar{X}_0$ だけではなく，出力 $\bar{Y}_0$ もまた入力になる．ここで，1 つの疑問が生まれる．いつ入力にフィードバックがあるのだろうか？　入力に出力が含まれるということは，どのようにしてすべての入力を確定させ，計算を始めるのだろうか？　これはまさにホップフィールドネットワークと同じ問題を抱えている．単一のニューロンを層に置き換えることで生じる．ここでもまた，同期型の方法で値を更新すれば良いだろう．つまり，最初のステップでは $\bar{X}_0$ だけで $\bar{Y}_0$ を計算する．次のステップでは，2 つの新たな入力（$\bar{X}_1$ と $\bar{Y}_0$）を得る．$\bar{X}_1$ と $\bar{Y}_0$ によって $\bar{Y}_1$ を得る．また，$\bar{Y}_2$ も同じように順次得ることができる．

---

7) 再帰型：recurrent，本書では再帰型と回帰型とは区別せず再帰型を用いる．

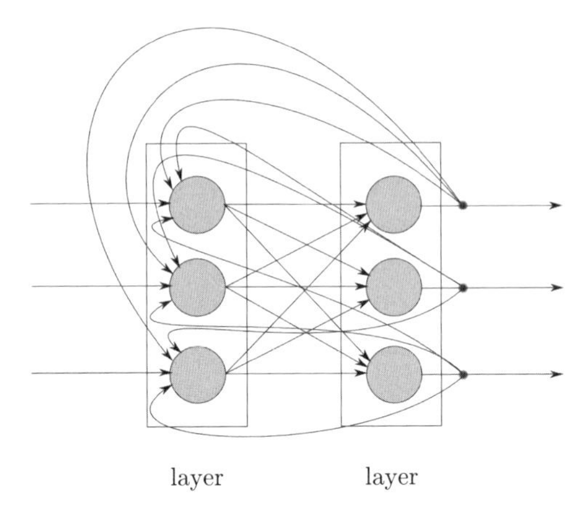

Fig. 6.2 An example structure of a two layers, $3 * 3$ recursive network (RNN). The two layers are fully connected like a standard ANN, but the node feeds the output back to the input. The feedback is delayed one time step, so these networks determine the output based on the current input and a memory of the previous one.

See figure 6.3 for a sketch of this recursive process. If you imagine that the series $\{\bar{X}_0, \bar{X}_1, \bar{X}_2, \dots\}$ is a time dependent signal, you understand that any single output $\bar{Y}_i \in \{\bar{Y}_0, \bar{Y}_1, \bar{Y}_2, \dots\}$ is influenced not only by $\bar{X}_i$ but also all the preceding $\bar{X}_{i-1}$,$\bar{X}_{i-2}$, etc. We say that the output has memory of the *history* of the signal $\bar{X}$.

Notice that in an RNN, neurons are not bi-stable like in a Hopfield network, but they can assume a continuous value between -1 and $+1$. So the network can represent more interesting physical signals than a Hopfield network. Moreover, in the latter network the weights are fixed and calculated with the Hebbian rule of equation 6.5. In a layered RNN instead, we have the advantage that we can *optimize* the Ws! We can proceed like in chapter 4 and do a training session with training sets $\bar{Y}_i^t \in \{\bar{Y}_0^t, \bar{Y}_1^t, \bar{Y}_2^t, \dots\}$. During this *learning* phase the error will be back-propagated and the Ws optimized. When the error will be small enough, the network will be ready to use.

You understand that the back-propagation procedure cannot be identical to that of chapter 4 or 5. Here we have signals that are fed back, so the

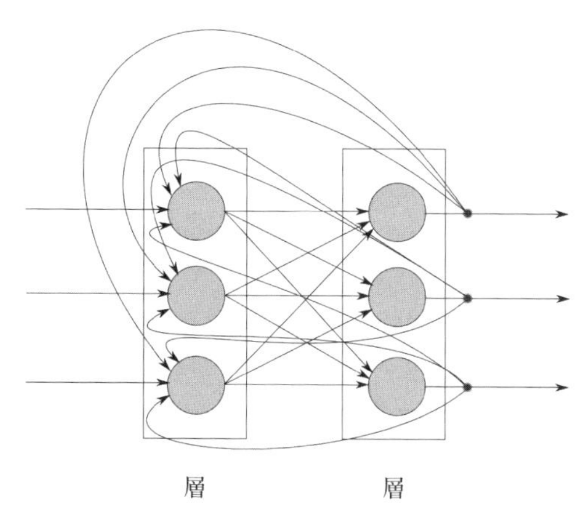

図 6.2　新規に 2 つの層からなる 3 × 3 の再帰型ニューラルネットワークを構築する (RNN). 2 つの層は，標準的な ANN と同様にすべてのニューロンは互いに結合されている．ANN との違いは，出力を入力に戻す結合，つまりフィードバックの存在である．フィードバックのタイミングは，時間ステップで遅れが生じる．つまり，RNN では，現在の入力と 1 つ前の記憶を基にした出力が得られる．

この再帰的な過程を図 6.3 に示す．

時間に依存した，連続の信号 $\{\bar{X}_0, \bar{X}_1, \bar{X}_2, \ldots\}$ を仮定すると，出力 $\bar{Y}_i \in \{\bar{Y}_0, \bar{Y}_1, \bar{Y}_2, \ldots\}$ は，$\bar{X}_i$ だけでなく，以前の $\bar{X}_{i-1}$ や，$\bar{X}_{i-2}$ 等に影響を受ける．つまり，その出力は，信号 X の履歴をもっている．

RNN に注目すると，各ニューロンはホップフィールドネットワークのような 2 状態のモデルではない．つまり，-1 から $+1$ の間の連続した値をもつ．これは，ホップフィールドネットワークよりも物理的な信号を再現する．さらに付け加えれば，後者のネットワークの結合荷重は，式 6.5 のヘッブの法則により修正され計算される．階層型の RNN では，結合荷重 W の最適化に有利にはたらくだろう！　すでに学んだ，第 4 章のような手順で教師データ $\bar{Y}_i^t \in \{\bar{Y}_0^t, \bar{Y}_1^t, \bar{Y}_2^t, \ldots\}$ による学習を進めることができる．学習段階では，誤差を伝播し，結合荷重 W が最適化される．誤差が十分小さくなると，ネットワークを使用する準備が完了する．

バックプロパゲーションの仕組みは，すでに第 4 章と第 5 章で学んだそれとは異なる．階層型の RNN では，信号に過去の情報も含まれることで，バックプロパゲーションのアルゴリズムはより複雑化する．ここでは詳細に触れ

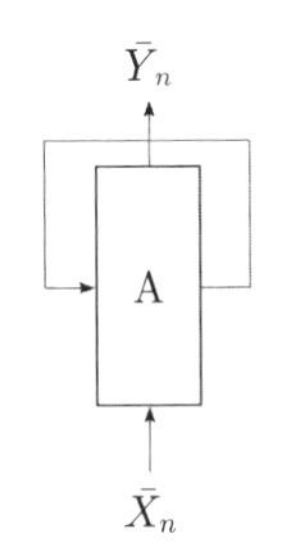

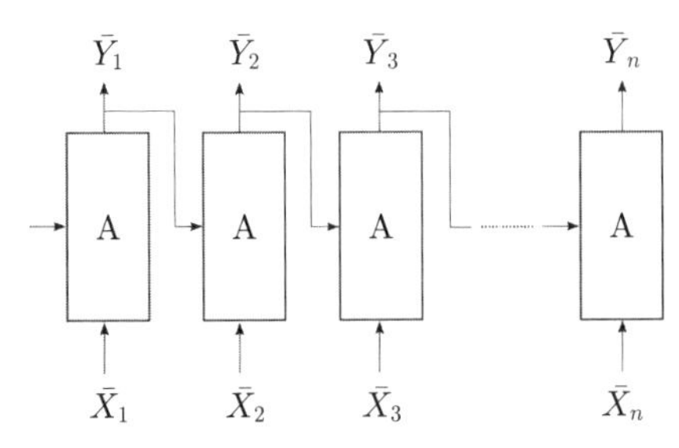

Fig. 6.3 Two ways to represent the same recursive network layer A. On the top the compact representation, on the bottom panel the time series $\{\bar{X}_1,\bar{X}_2, \bar{X}_3 \ldots\}$ is expanded. The multiple input $\bar{X}_1$ is transformed by the block A and forms output $\bar{Y}_1$, and this will be the next input together with $\bar{X}_2$ and so on).

back-propagation algorithm must be more complex. We do not want to go in the details here, but please remember that modern neural network libraries like Keras or other have these optimization procedures built in. You can use them even without understanding the details about how they work.

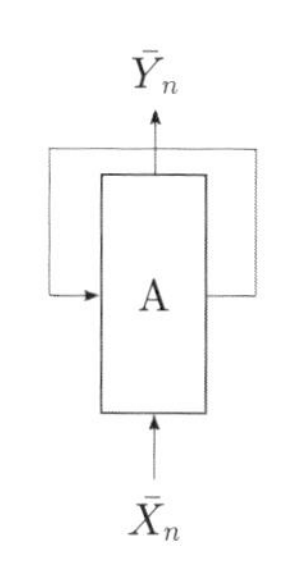

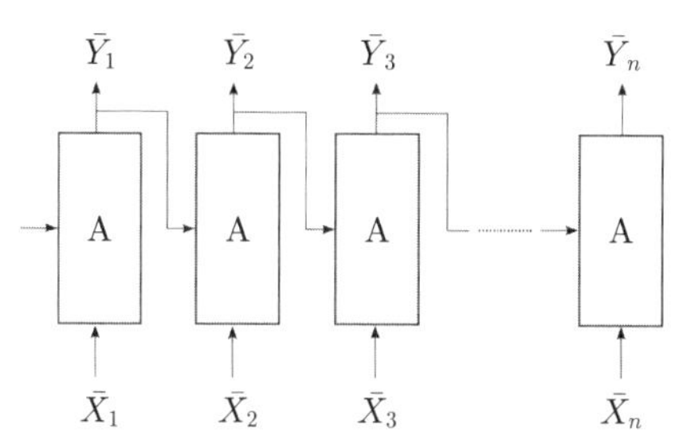

図 6.3　A 層をもつ再帰型ニューラルネットワークを 2 つの方法で示す．上側のコンパクトな図は，下側の時間的な連鎖 $\{\bar{X}_1, \bar{X}_2, \bar{X}_3 \ldots\}$ によって展開される．複数入力 $\bar{X}_1$ は，ブロック A によって変換され，$\bar{Y}_1$ を出力する．また，出力 $\bar{Y}_1$ は，$\bar{X}_2$ とともに次の入力となる．その後，これらの処理が繰り返される．

ないが，最新のニューラルネットワークライブラリーの Keras 等を用いれば，難しい設定は抜きにして，ネットワーク最適化の手順を体験することができるだろう．

Think about these questions

- What is the *field* in a Hopfield network?
- What is the general equation regulating a Hopfield network?
- What is the difference between a Hopfield network and a perceptron?
- What is the *Hebbian rule*?
- Why do we say that a Hopfield network has memory?
- What is the difference between a *synchronous* and *asynchronous* network?
- What is the *crosstalk* of a Hopfield network?
- What do we have to do to increase the *capacity* of a Hopfield network?
- What are the advantages of a RNN network against a Hopfield one?
- Make a program that calculates the Hopfield network output in an *asynchronous* way.

考えてみよう

- ホップフィールドネットワークにおける場とはどのようなものか？
- ホップフィールドネットワークを最適化する一般式はどのように記述されるか？
- ホップフィールドネットワークとパーセプトロンの違いはどのようなことか？
- ヘッブの法則とはどのようなものか？
- ホップフィールドネットワークが記憶をもっていると言える理由はどのようなことか？
- 同期型のネットワークと非同期型のネットワークの違いはどのようなことか？
- ホップフィールドネットワークの相互干渉とはどのようなものか？
- ホップフィールドネットワークで容量を増やすためにはどうすれば良いのか？
- ホップフィールドネットワークに対して，再帰型ニューラルネットワークの利点はどのようなことか？
- ホップフィールドネットワークの出力を非同期型の方法で計算するプログラムを組んでみよう！

Chapter 7

Visual perception

At the end of 1800s a group of experimental psychologists in Germany were able to identify several visual perception rules by which humans recognize and group objects together. These rules were described using a number of principles that are now generally called the *Gestalt theory*, from the German word meaning *shape* or *form*. The Gestalt establishes how the human visual system processes stimuli together to form meaningful *scenes*. A visual scene is firstly perceived by its *pre-attentive* elements. With the word pre-attentive, we mean elements that have no particular meaning by themselves. Those could be simple parts of objects like edges, corners, curves or particular *features* like the color, movement, texture, luminosity etc. These parts are processed accordingly to the Gestalt rules. As a result, objects are recognized and grouped in a meaningful way to form a visual scene. On this scene, human *attention* can operate: some object may attract our attention or we may choose to attend to or *search* for some others. On the other hand, we are not able to control or alter the pre-attentive visual elaboration steps. The Gestalt theory is about the principles that govern these unconscious pre-attentive elaboration phases. Hereafter we describe six well known Gestalt principles by use of simple examples.

7.1 The six Gestalt principles

7.1.1 Proximity

Our visual perception system has the ability to group things together very efficiently. The proximity principle says that objects may be perceived as a single one, or as a group, if they are near each other. For example, see

第7章

視覚認知

1800年代末，ドイツの心理学者のグループは人間が物体を認識して分類する際の視覚認知に関して，いくつかの規則を実証した．これらの規則は，次のようないくつかの原則により説明される．形や形式を意味するドイツ語の単語から，現在は一般的にゲシュタルト理論と呼ばれている．ゲシュタルト理論は，人間の視覚システムが意味のある場面を形成するために，どのようにして信号刺激を作っていくのかを立証する．まず視覚的な場面は，対象物に対する注意以前の要素によって認識される．注意以前とは，私たちにとって要素それ自体では特に意味がないものである．それらは，ふちや角，曲線のように対象物の単純な要素，あるいは色，動き，質感，光度といったような特定の特徴であることもある．これらの要素は，ゲシュタルトの法則に従って処理される．結果として，物体は視覚的な場面を形成するために意味のある方法で認識され，分類される．場面によっては人間の注意が作用することがある．つまり，ある対象物が私たちの注意を引くかもしれないし，あるいは私たちが他の何かに注目したり，他の何かを探すことを選ぶかもしれない．その一方で，注意以前の視覚が作っていく過程を制御したり，変更することはできない．ゲシュタルト理論とは，これら無意識の注意以前の認識過程を支配する原則である．次の節では，簡単な例を用いてよく知られたゲシュタルトの6つの原則を説明する．

7.1 ゲシュタルトの6つの原則

7.1.1 近接性

私たちの視覚システムは，物事を非常に効率的にまとめることができる．近接性は，物体が互いに接近している場合，それらは単一のものとして知覚されることを意味する．図7.1に示すように，同一の物体がランダムな距離に配置

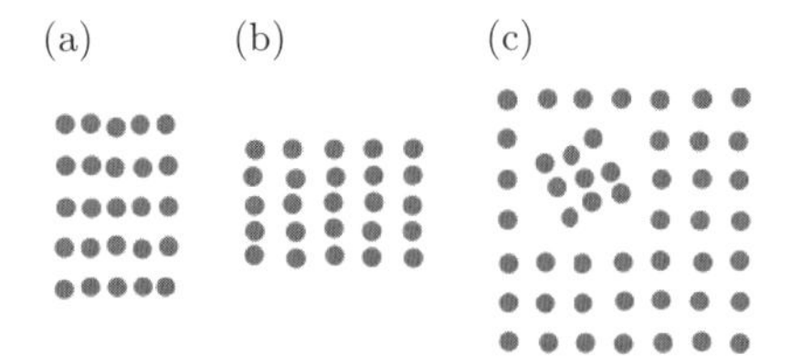

Fig. 7.1 Proximity principle: in panel (a) it looks as if it is made of rows of circles, but in (b) it looks as if it is made of columns of circles due to grouping by proximity. In (c) our perception system focuses our attention to the 9 items in the top left. Those are felt as an independent group, just because the items are closer to each other.

figure 7.1 for a simple sketch where identical objects are placed at random distances. We perceive vertical, horizontal lines or even groups depending on the relative distances between items. In panel (c) for example, some objects are nearer to each other than the others. As a result, they appear as a distinct *group*. Notice that distances are random and there is not a real association or relation between items. The mere fact that their average relative distance is arranged in a particular way, creates these perceptions of independent groups of objects.

7.1.2 Similarity

Similarity between items in a scene is very important for our visual system to organize objects in groups. The principle of similarity can be understood by the example in figure 7.2. In panel (a) we see a number of identical blocks arranged randomly. Eight of them are placed along a circle, however it is not easy to spot the circle. But if we substitute these blocks with a different shape, we have the clear perception of the circle (panel (b)). The human visual system has the ability to group things by their similarity, that's why the circle is now very easy to spot.

7.1.3 Common fate principle

This principle is related to movement: if two or more items move together,

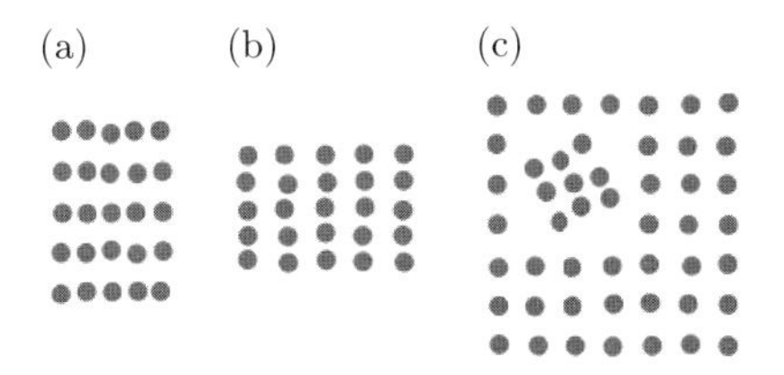

図 7.1 近接性：パネル (a) では，行区切りで丸が並んでいるように見えるが，パネル (b) では列区切りで並んでいるように見える．これが近接性によるグループ分けの作用である．またパネル (c) では左上の9つの丸に注意が向く．それらが独立であるように感じる理由は，9つの丸の間隔はランダムだが，他の間隔に比べてより近いことによる．

されている場合を考えよう．物体間の距離によって，垂直な線，水平な線，または集合体を認識することができる．例えばパネル (c) では，いくつかのオブジェクトが他のオブジェクトよりも近くにある．結果として，それらは異なるグループとして識別される．実際上，それらの関係性に意味はない．つまり，平均の相対距離が他のものより短いという事実だけで，独立したオブジェクトのグループであると識別される．

7.1.2 類似性

それぞれの場面におけるアイテム間の類似性は，物体をグループにまとめるための視覚システムにとって非常に重要な役割を担う．類似性の原則は，図 7.2 の例で理解できる．パネル (a) では，ランダムに配置された多数の同一ブロックが見える．そのうち8個は円に沿って配置されているが，円を見つけるのは容易ではない．しかし，これらのブロックを別の形に置き換えると，円をはっきりと認識できる（パネル (b)）．人間の視覚システムには，類似性によって物事をグループ化する機能がある．そのため，パネル (b) では円は非常に見つけやすくなっている．

7.1.3 共通概念の原則

次に，動きに関連した共通概念について述べる．2つ以上のアイテムが一緒

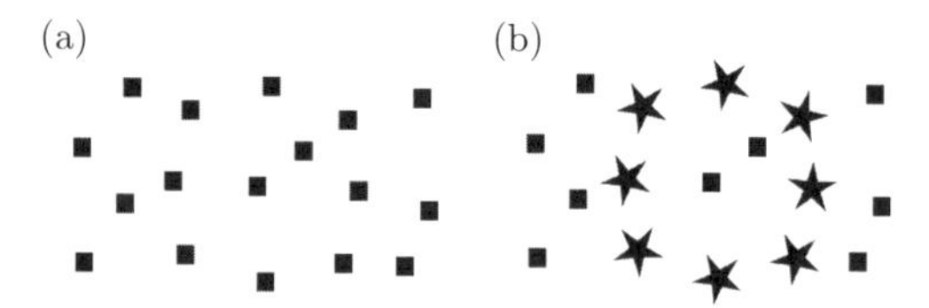

Fig. 7.2 In panel (a) you perceive randomly placed identical items. In panel (b) we see a group of stars along a circle. The positions of the stars correspond exactly to the positions of the items in panel (a). The circle emerges in our perception only in (b) because all the stars are *similar* to each other.

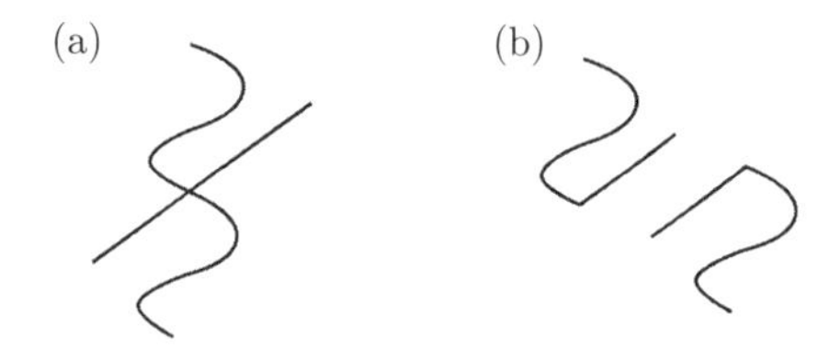

Fig. 7.3 If panel (a) is shown, a subject prefers to recognize a single sinusoidal curve and a straight line. However, the image could be interpreted as well as the two objects in (b) placed adjacent to each other. Our brain strongly prefers the smoother interpretation of reality.

and if their motion is identical, they are perceived as a group. Suppose you have an image where there are many similar objects, like in previous figure 7.1. If you imagine that some of those move together, they will be perceived as a group. You can imagine that one of the circles in the central group of figure 7.1 (c) and another circle in the outside are moving together, those two will be perceived as a single moving object. This happens in contradiction with the proximity principle that should assign those objects to different groups. This shows that the common fate principle is very strong and dominant over other principles.

7.1.4 Continuity

When we perceive a curve the visual system always identifies objects accordingly to *continuity* rules. For example, in figure 7.3 (a) we see a sinusoidal curve crossing over a straight line. However, the image could be

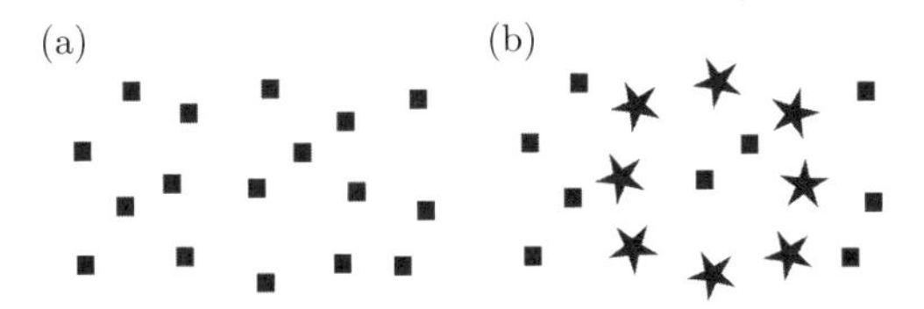

図 7.2　パネル (a) では，ランダムに配置された丸が複数見える．パネル (b) では，円に沿っている星のグループが見える．星の位置は，パネル (a) の丸の位置と正確に一致している．すべての星は互いに類似性をもつので，円は (b) でのみ認識される．

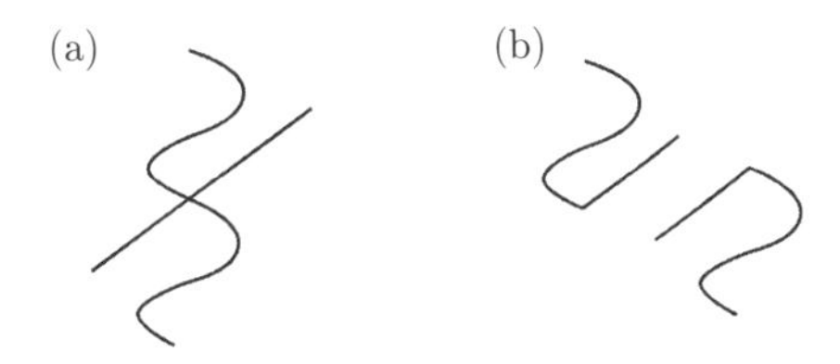

図 7.3　パネル (a) が表示されている場合，被験者は単一の正弦曲線と直線を認識する傾向にある．ただし，(b) の 2 つのオブジェクトが互いに隣接して配置されている場合と同様に画像は解釈できる．私たちの脳は，現実的によりスムーズな解釈をより好む傾向がある．

に動く場合，そしてそれらの動きが同一である場合，それらはグループとして認識される．図 7.1 のように，似たような物体がたくさんある画像があるとする．それらのうちのいくつかが一緒に動くとき，それらはグループとして知覚されるだろう．そうすると，図 7.1(c) の中央のグループの円の 1 つと別の外側の円が一緒に動いていると想像することができる．それら 2 つは単一の動いている物体として知覚されるだろう．これは，それらの物体を異なるグループに割り当てるべき近接原理と矛盾している．つまり，共通概念の原則が他の原則よりも非常に強く優位にはたらくことを示している．

7.1.4　連続性

曲線を知覚するとき，視覚的なシステムは常に連続性の規則に従って物体を識別する．たとえば図 7.3 (a) では，正弦曲線が直線と交差している．しかし，イメージは，(b) のように，互いに隣接して配置されている 2 つのオブジ

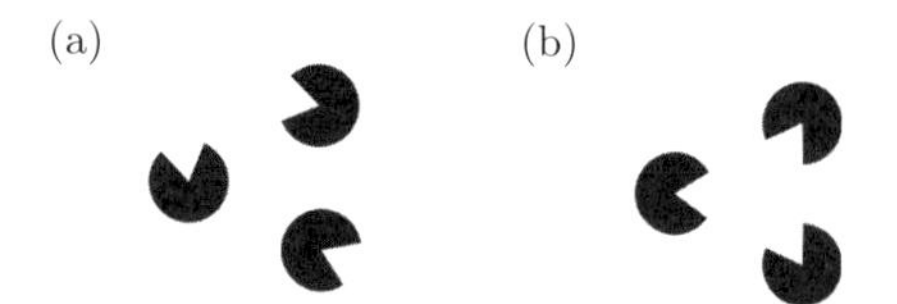

Fig. 7.4 Emergence principle: In (a) we see three *pacman* shaped objects randomly oriented, whereas in (b) the orientation is aligned in a particular way. The shape of a triangle *emerges* from the image. However, no triangle is present at all in (a) or (b). These images from the information point of view, are identical.

formed by the two objects in panel (b) placed adjacent to each other. Our brain always prefers to choose the smoothest and continuous interpretation of reality.

7.1.5 Emergence or Closure

The visual perception of human and animals always tries to recognize shapes and objects from past visual experiences. This is true to the level that even nonexistent or incomplete objects can be perceived. For example look at figure 7.4 (a) and (b). Those two stimuli are geometrically equivalent and do contain the same amount of information. However, the one in (b) shows the illusory *emergence* of a triangle. This principle is also called *closure*, meaning that the brain has the ability to complete or close an open form if there are some of its parts. In the example of the figure, parts of a triangle can be seen and the brain creates the perception of a complete one.

7.1.6 Figure-ground

When we observe a scene, the brain tries to isolate an object from the background. In many cases there is not enough information available to clearly separate background from foreground objects. In figure 7.5 we give few examples where our perception system has difficulties in definitely distinguishing a figure from background.

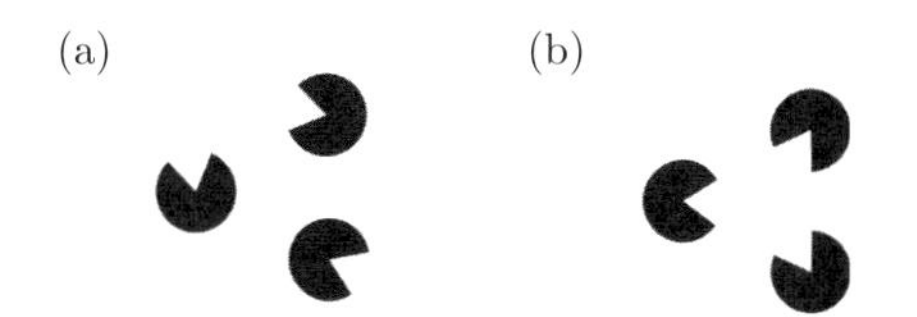

図 7.4 出現の原則：(a) では，3 人のパックマンの形がランダムな方向におかれたオブジェクトが見える．(b) では，方向は特定の方法で整列されており，画像から三角形の形状が浮かび上がってくる．しかしながら，(a) または (b) の中に三角形はまったく存在しない．情報の観点から見たこれらの画像は同一である．

ェクトによって形成されている可能性がある．私たちの脳は常に現実の最も連続する解釈を選択しようとする．

7.1.5 出現または閉鎖

人間や動物の視覚的認識は，過去の視覚的経験から常に形や物を認識しようとしている．これは，存在しない物体や不完全な物体でも認識できるということである．例として，図 7.4 の (a) と (b) を示した．

これら 2 つの刺激は幾何学的に等価であり，同じ量の情報を含む．しかし，(b) は三角形の錯覚的な出現を示している．この原則は閉鎖（クロージャ）とも呼ばれる．脳にその情報の一部がある場合に，私たちはある形の出現を認識することができる．あるいは，閉鎖する作用があることを意味する．図の例では，三角形の一部が見え，脳は完全なものの知覚を生み出している．

7.1.6 図地（フィギアグラウンド）

私たちがある場面を観察するとき，脳は物体を背景から分離しようとする．多くの場合，背景と前景のオブジェクトを明確に区別するのに十分な情報はない．図 7.5 では，知覚システムが物体と背景を明確に区別することが困難である例をいくつか挙げている．

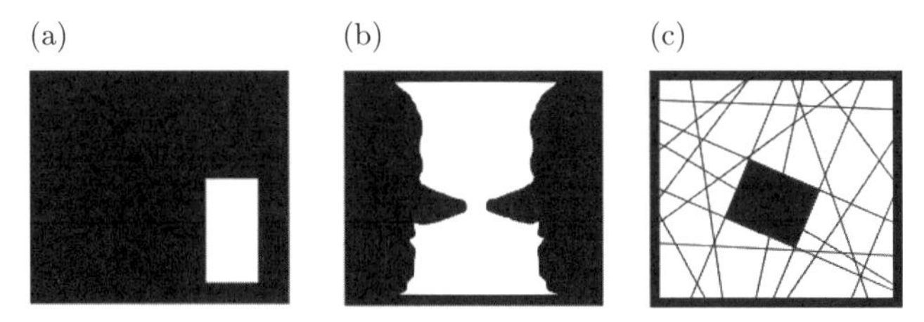

Fig. 7.5 Do you perceive the foreground in these panels correctly? In (a) do you see a dark wall with an open door on a white background, or do you see a white rectangular box in a dark room? In (b) do you see an old-style white cup? Or do you see the faces of two men looking at each other in a white background? Do you see in (c) a network of lines with a square opening over a dark background or the opposite?

7.2 The Fourier space

The *Fourier transformation* is of paramount importance, and in several cases, we perceive sensory data in the Fourier space. What do we mean by this, and what is the Fourier space?

First of all, let's intuitively explain what Fourier transformation is. It is an operation by which a time dependent function is transformed into a frequency dependent function. In practice, if our time dependent signal $f(t)$ (it could be a sound for example) has some periodicity, its Fourier transformation is a frequency dependent function where the periodic components of the signal are represented[1]. Mathematically speaking, the *Fourier transformation* is expressed with this formula:

$$F(\omega) = \int_{-\infty}^{\infty} f(t)e^{-i\omega t}dt \qquad (7.1)$$

To understand the meaning of it, let's remember that the term $e^{i\omega t}$ is $cos(\omega t) + i * sin(\omega t)$. So, to simplify, let's choose $\omega = \omega_0$ and forget the imaginary part. We write again the integral:

1) Remember that the period ω is $2\pi f$ where f is the frequency.

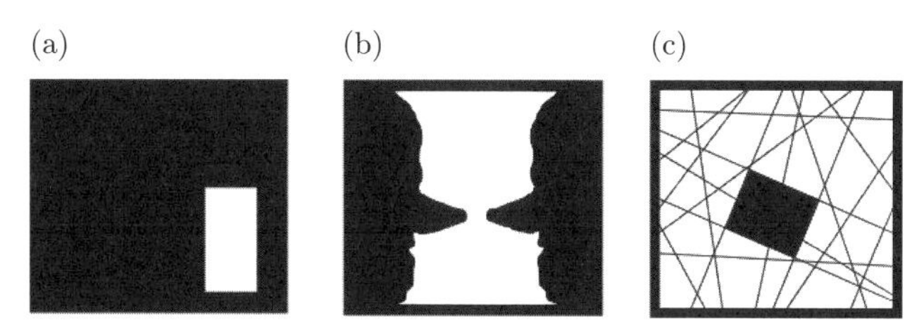

図 7.5　左から順にパネルの前景を正しく認識できているだろうか？　(a) は，白い背景の上に開いたドアがある黒い壁が見えるか？　それとも暗い部屋の中に白い長方形の箱が見えるか？　(b) では，古いスタイルの白いコップが見えるか？　あるいは，白い背景の中に互いを見ている 2 人の男の顔が見えるか？　(c) では，暗い背景の上に正方形の開口部をもつ線のネットワークが見えるか？　それともその反対が見えるか？

7.2　フーリエ空間

フーリエ変換は現実の知覚システムを理解するうえで最も重要になる．なぜなら多くの場合，私たちはフーリエ空間を通じて情報を抽出するからである．これはどういう意味だろうか？　フーリエ空間とは何だろうか．その前にまず，フーリエ変換とは何かを直感的に説明する．フーリエ変換は，時間依存関数を周波数依存関数に変換する操作である．実際には，時間依存信号 $f(t)$（例えば音）はいくらかの周期性をもっている．フーリエ変換は信号の周期的な成分が表される周波数依存関数である[1)]．数学的表現を用いれば，フーリエ変換は次式で表される．

$$F(\omega) = \int_{-\infty}^{\infty} f(t) e^{-i\omega t} dt \tag{7.1}$$

その意味を理解するために $e^{i\omega t}$ が $cos(\omega t) + i * sin(\omega t)$ であることを覚えておこう．式を簡単にするために，$\omega = \omega_0$ にして，虚数部を消去する．そして，もう一度積分する．

$$F(\omega_0) = \int_{-\infty}^{\infty} f(t) sin(\omega_0 t) dt \tag{7.2}$$

$f(t)$ が $\omega = \omega_0$ で同じ正弦波だとすると，積分は $\int_{-\infty}^{\infty} sin^2(\omega_0 t)dt$ である．

1) 周波数が f である時，周期 ω は，$2\pi f$ であることを思い出そう．

$$F(\omega_0) = \int_{-\infty}^{\infty} f(t) sin(\omega_0 t) dt \tag{7.2}$$

If we assume that f(t) is a sine wave with exactly the same $\omega = \omega_0$, then the integral is $\int_{-\infty}^{\infty} sin^2(\omega_0 t)dt$. This represents the area under the curve $sin^2(\omega_0 t)$.

Because of the sine wave square the area is always positive and the result is not zero. However, if ω is different from ω_0, the integral $\int_{-\infty}^{\infty} sin(\omega t) sin(\omega_0 t)dt$ will have negative and positive parts. Integrating those over a wide interval will result to a total of zero.

So, the Fourier transform gives a number different from zero only when the function $f(t) = sin(\omega t)$ has exactly the same frequency ω_0. The graph of $F(\omega)$ will look like a straight line on zero, with a single peak at $\omega = \omega_0$. Now, this is fine if the function $f(t)$ is a sine wave, but it is also true for the general case of any function f(t). If the f(t) is a complex sound, see figure 7.6, you can understand that the Fourier transformation emphasizes all the sinusoidal components of the signal. There will be peaks corresponding to all the frequencies of that sound. For example, when we hear the sound of a piano, can you tell if one, two or three notes are played? Yes, you can. You can immediately and easily tell that. Why? Because acoustic perception lives in the *Fourier space*! We only perceive frequencies, we are conscious of the two peaks of the Fourier spectrum, but we are not aware of the time dependence of the sounds (see figure 7.6 (b)).

Of course, we also have senses other than the sense of hearing. We also see, touch, taste and do other things. The Fourier transformation is very important for the other senses, too. For the sense of touch, the frequency of the tactile input is very important. If you remember chapter 1, what we feel is strongly related to the frequency of the vibration of what we touch. This dependence of our senses to frequencies is also noticeable in vision. For example, which do you notice more, a blinking light or a steady one? Of course the blinking one. And the frequency of blinking will influence how

これは，曲線 $sin^2(\omega_0 t)$ の下の面積を表す[2)].

曲線は正弦波の2乗であるため，その面積は常にプラスであり，結果はゼロにはならない．ただし，$\omega = \omega_0$ と異なる場合，積分値 $\int_{-\infty}^{\infty} sin(\omega t) sin(\omega_0 t) dt$ は正の値と負の値になる．広い範囲でそれらを足し合わせると結果はゼロになる．

したがって，フーリエ変換は，関数が $f(t) = sin(\omega t)$ で，厳密に同じ周波数 ω_0 をもつ場合に限り，ゼロとは異なる数を与える．$F(\omega)$ のグラフは，$\omega = \omega_0$ に単一のピークをもつ，数値ゼロの直線のように見える．関数 $f(t)$ が正弦波であればこれでもちろん問題ないが，一般的な関数 $f(t)$ の場合にも当てはまる．$f(t)$ が図7.6に示すように複雑な音の場合でも，フーリエ変換をすれば信号に含まれている正弦波成分をが明らかになり，音のすべての周波数に対応するピークが現れる．例えば，ピアノの音が聞こえたら，1つ，2つ，または3つの音が演奏されているかどうかわかるだろうか？　もちろん，簡単にわかり，答えることができる．これは，音響的な認識はフーリエ空間にあるからである．私たちは周波数を知覚するだけで，フーリエスペクトルの2つのピークを意識している．しかし，音の時間に対する依存性は認識できていない（図7.6 (b)）.

もちろん，聴覚だけがあるわけではない．私たちは見たり，触ったり，味わったり，他の知覚を行う．他の感覚にとってもフーリエ変換は非常に重要である．触覚にとって，触覚的な入力の周波数は非常に重要である．第1章で述べたように，私たちが感じるものは私たちが触れたものの振動の周波数に強く関連している．この感覚の周波数への依存は，視覚でも顕著である．例えば，点滅する光と，点滅しない光とではどちらにより気がつくか？　もちろん点滅しているほうである．そして点滅頻度はその効果がどれほど目立つかに影響する．そのため，灯台，救急車，パトカーは光が点滅するように作られている．フーリエ変換は，時間領域だけでの知覚で重要であると考えるかもしれない．しかし，もっと一般的な規則がある．空間フーリエ変換は視覚において，その役割を果たしている．私たちは物体の空間的な位置を非常に良く認識し，特定

2) 周期 ω は，$2\pi f$ であり，f は周波数である．

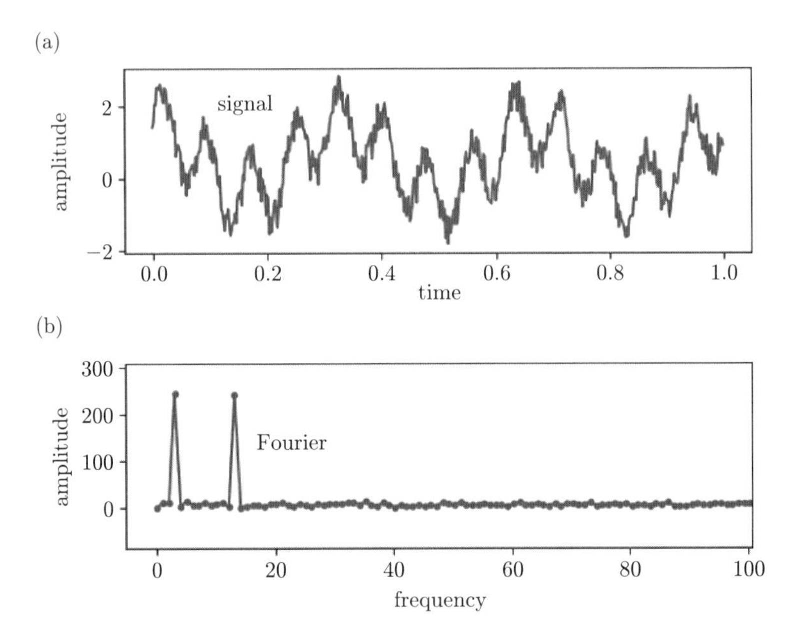

Fig. 7.6 The spectrum of a sound signal. In (a) the signal is plotted in function of time. We can recognize that there are two main frequencies, two slower ones of large amplitude and a faster wave of smaller amplitude (noise). In the other panel (b) the Fourier transformation of this signal is shown: the value is always close to zero but for the two frequencies contained in (a). The horizontal axis is time in panel (a), whereas it is the frequency in panel (b).

prominent the effect is. That's why we make lighthouses, ambulances and police cars with blinking lights. You may think that Fourier transformation is important in perception only in the time domain. But there is a more general rule. The spatial Fourier transformation has a role in vision. Even though we are very sensitive to the spatial positions of objects and can easily direct attention to a specific spatial location, studies show how the visual system is influenced by spatial frequencies. For example, we tend to notice certain spatial frequencies more than others[Read, 2019].

7.3 Visual search

A very important method to understand the brain functioning (in par-

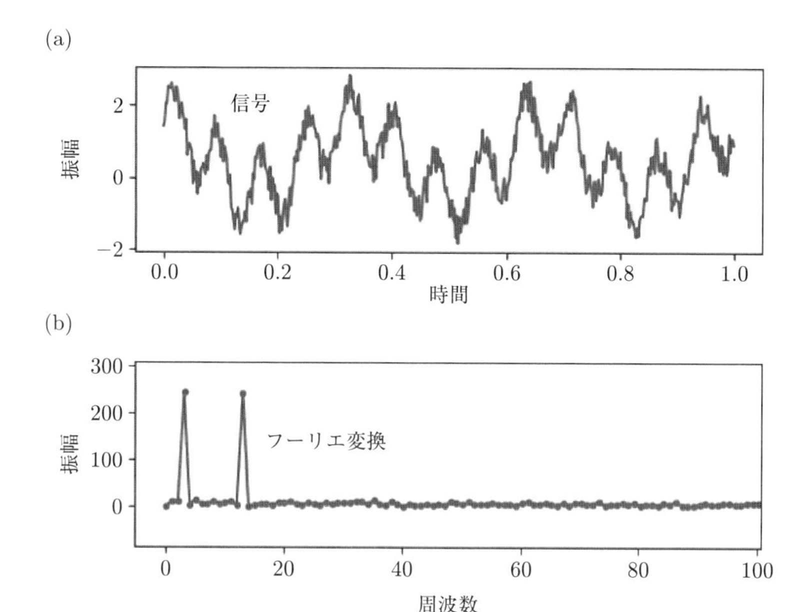

図 7.6　音声信号のスペクトルの例．(a) では時間の関数で信号がプロットされている．2 つのおもな周波数，より高い振幅のより遅いものとより速いものがあることを認識できる．(b) では，この信号のフーリエ変換が示されている．値は常にゼロに近くなるが，(a) に含まれる 2 つの周波数についてである．横軸は (a) では時間であり，(b) では周波数である．

の位置に直接注意を向けることができるが，これらの視覚システムが空間周波数によってどのように影響を受けているのかについて多くの研究がある．例えば，特定の空間周波数に気づく傾向がある [Read, 2019].

7.3　視覚的探索

脳の機能（特に人間の視覚）を理解するための非常に重要な方法は，視覚探

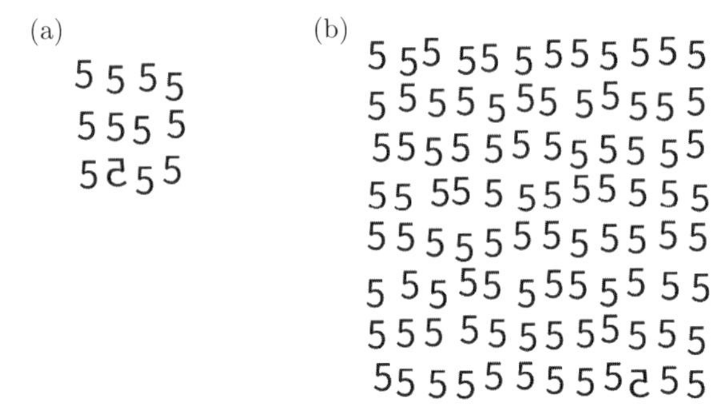

Fig. 7.7 A difficult visual task: search for the inverted "5". In panel (a) there are 12 items, in panel (b) there are 120 items. You notice that it takes longer to find the target when there are more items.

ticular human vision) is to study visual search. This is an experimental methodology in which a subject is presented a complex pattern with many items, and he/she is asked to search for one of them (the target). To understand what we mean, please see figure 7.7.

In panel 7.7 (a), please search for the inverted character "5". You will notice that there are so many "5"s and so it is not easy to search, especially in panel (b) where there are much more distractors (the "5"s). The important thing in this task is that we can actually measure the difficulty quantitatively, by measuring the *reaction time*! So let's imagine a very simple experiment. We use a computer to display many panels with a variable number of items. In our case, the *distractors* are the character "5" and the *target* is the inverted character "5". We ask the subject to click a key if it is present on the screen. In this experiment there will be only one target and a variable number of distractors (in half of the tests there will be no target, only distractors). If you run this experiment[2)] and plot the number of items on the horizontal axis and the reaction time on the vertical one, you will obtain a linear dependence like the one in figure 7.8.

These results are perfectly consistent with our eyes *scanning* the visual

2) A public dataset of visual search "2"s among "5"s is available from Jeremy Wolfe visual attention laboratory[Wolfe]

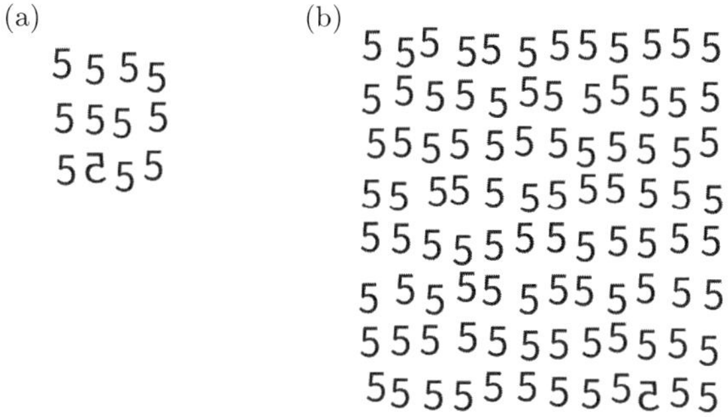

(a)　(b)

図 7.7　難しい視覚的作業の例：逆さの “5” を探す．パネル (a) では 12 個，(b) では 120 個の項目がある．項目が多いほどターゲットの検索に時間がかかることがわかる．

索を研究することである．これは実験的な方法論であり，そこでは被験者は多くの項目を含む複雑なパターンを提示され，それらのうちの 1 つ（ターゲット）を探すように求められる．図 7.7 を見てほしい．

図 7.7 のパネル (a) で “5” の反転文字を探索してみよう．“5” が非常に多いので，探索は簡単ではない．特にパネル（b）の場合は，より多くのディストラクタ (“5”) がある．この作業で重要なことは，反応時間を測定することによって定量的な測定が行えること，つまりその課題の困難さを測定できることである．それでは，非常に単純な実験を想像してみよう．コンピュータを使って様々な数の項目をもつ多くのパネルを表示している．この場合，ディストラクタは文字 “5” で，ターゲットは反転文字 “5” である．被験者にはターゲットが画面上にある場合は，キーボードをクリックするように促す．この実験ではターゲットは 1 つと可変数のディストラクタしかない（テストの半分にはターゲットはなく，ディストラクタだけがある）．この実験を実行した後[3]，横軸に項目数，縦軸に反応時間をプロットすると，図 7.8 のような線形関係が得られる．

これらの結果は目が視覚入力をスキャンするのと完全に一致している．実際には，図の線の傾きは 1 項目あたりのスキャンにかかる時間を表しており，それはミリ秒単位で測定される．それは研究者に，視覚システムの注意が単一

3) 5 の中に 2 を探す実験に関するデータは Jeremy Wolfe visual attention laboratory より利用可能である．

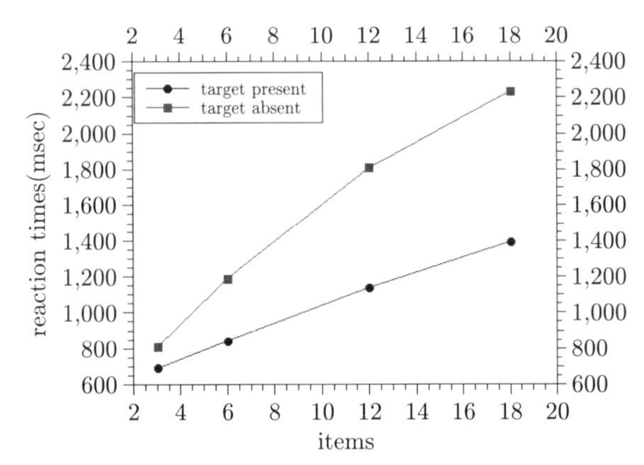

Fig. 7.8 Reaction times collected in an experiment of visual search[Wolfe]. Notice the linear dependence on the number of items and the fact that the inclination of the curve is steeper when the target is absent (squares).

input. Actually the inclination of the line in the figure is measured in milliseconds per item. It gives the researcher an information on how long, on average, the attention of the visual system stays on a single item. Noticeably, when the sample is missing, the reaction time is about twice as much time. This is exactly what we expect in the case of scanning. In fact, if the target is missing, we have to examine all the items before we can conclude that the target is missing. On the other hand, if the target is present, and placed in a random position, it will take on average half the time before it is found. Can we conclude that our visual system is just a robot that scans through items? When we search something, do we just scan all along the visual pattern until the target is found? It is possible, but there are experiments that contradict this hypothesis.

See an example of visual search in figure 7.9. Now we have many circles and a plus. It is very easy to find the plus, our visual system recognizes the plus instantly, without the need of any effort. The time to detect the target is very short, and does not depend at all on the total number of items. The target just *pops out* the screen regardless of how many distractors are present. If you ponder on it, the visual search problem of figure 7.7 and

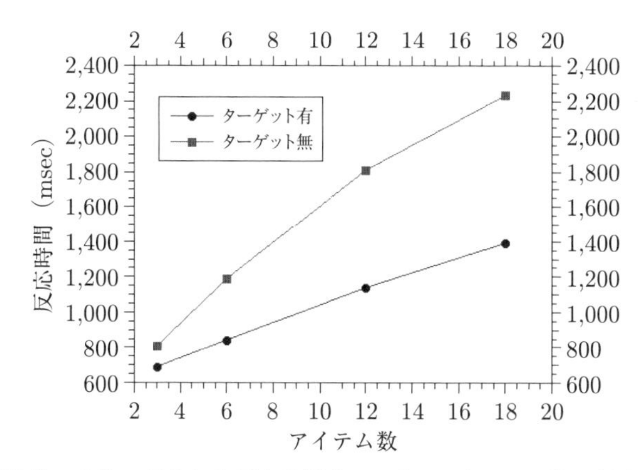

図 7.8　視覚探索の実験で収集した反応時間 [Wolfe]．アイテムの数に対する線形の依存関係と，ターゲットが存在しない場合の曲線の傾きが急であることに注目してみよう（四角）．

のアイテムにとどまるのにかかる時間の平均についての情報を与える．注目すべきことに，ターゲットがない場合に要する時間は，ターゲットがあるときに比べて約 2 倍である．これはまさに視覚システムがスキャンしていることを示している．実際，ターゲットが見つからない場合は，ターゲットが見つからないと判断する前にすべての項目を確認する必要がある．代わりにターゲットが存在していて，それがランダムな位置に置かれている場合，ターゲットが見つかるまでに平均で半分の時間がかかる．私たちの視覚システムは，アイテムをスキャンするロボットにすぎないと結論づけることができるだろうか？　何かを探索するときに，ターゲットが見つかるまで視覚パターン全体をスキャンするだろうか？　それは可能かもしれないが，この仮説と矛盾する実験結果がある．

例えば図 7.9 の視覚探索の例を参照してみよう．多数の◯と＋がある．＋を見つけるのは非常に簡単である．私達の視覚システムは努力しなくても，即座に＋を認識する．ターゲットを検出するのにかかる時間は非常に短く，アイテムの総数にはまったく依存しない．ターゲットは，ディストラクタの数に関係なく，画面から飛び出すように見つかる．よく考えてみると，図 7.7 と図 7.9 の視覚的探索問題は同等である．数学的および幾何学的な観点から見ると，これら 2 つの問題は同等だが，私たち人間にとっては，これらは非常に

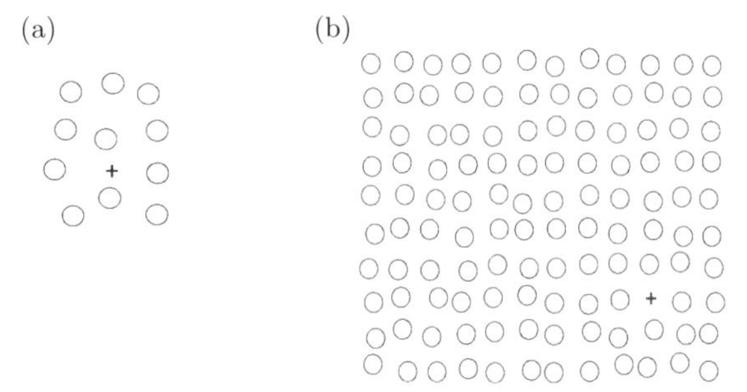

Fig. 7.9 Reaction times for *easy* visual search. Notice how easy it is to find the "plus" shaped target in this case compared to the case of figure 7.7. Reaction time is shorter and it doesn't depend on the number of items on screen. Nevertheless, the search problem is theoretically identical from the one in figure 7.7. Why can human brain solve this one easily but has difficulties with the other?

figure 7.9 are equivalent. From the mathematical and geometrical point of view, those two problems are equivalent. However, for us humans, they appear so different: one very difficult, the other very easy. Why is that?

In 1980s Treisman developed the Features Integration Theory [Treisman and Gelade, 1980]. According to this theory, when a stimulus is perceived the visual system firstly processes all the *features* of the image. For features we mean the colors, edges, corners, curvature, movements etc. These features are processes in the early stages of the perception in parallel. This early stage or processing is called *pre-attentive*. Later, the features are integrated and a conscious object is actually perceived in the so called *attentive* stage of visual perception. So why do we find some images so easy to recognize and some other not? According to the features integration theory, this happens because our brain pre-attentive processes classify images by their features and create a *feature map* somewhere in our brain. Fives and inverted fives have similar features (the corners, edges, curvatures) whereas a circle and a cross have very different ones. So it is very easy for our brain to find a cross in the middle of many circles, than finding an inverted five in the middle of normal ones. See figure 7.10 as an example of a schematic representation of such feature map in our brain.

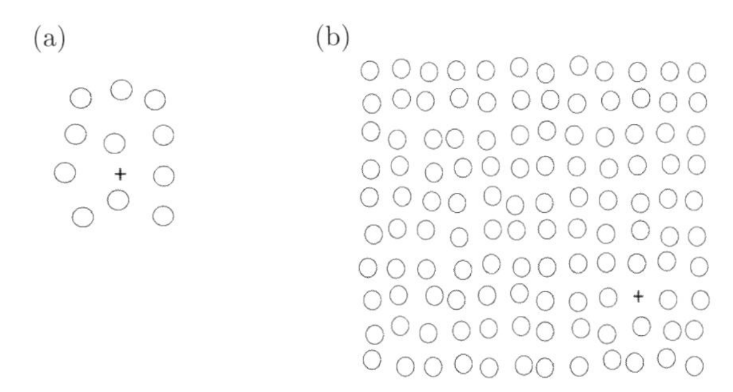

図 7.9　(a) 視覚探索が簡単な例．この場合，図 7.7 の場合と比較して，+ の形をしたターゲットを見つけるのは簡単である．つまり，反応時間が短く，画面上のアイテム数には左右されない．それにもかかわらず，検索問題は理論的には図 7.7 の問題と同じである．なぜ人間の脳はこれを簡単に解決できるのに，他の課題は難しいと感じるのだろうか．

異なるように見える．1 つは非常に困難で，もう 1 つは非常に簡単である．なぜだろうか？

1980 年代にトライズマンは特徴統合理論を開発した [Treismann and Gelade, 1980]．この理論によると，視覚的な刺激が到達すると，視覚システムは最初に画像のすべての特徴を処理する．特徴とは，色やふち，角，曲率，動きなどを意味する．これらの特徴は，知覚の初期段階におけるプロセスである．この初期段階または処理は，注意以前と呼ばれる．後に機能は統合され，意識的な物体は実際には視覚認識のいわゆる注意深い段階で認識される．それでは，なぜ認識しやすい画像と，そうでない画像があると感じるのだろうか．特徴統合理論によると，これは脳の注意以前のプロセスが画像をその特徴によって分類し，脳のどこかに特徴マップを作成するために起こる．5 と逆にした 5 は似たような特徴（角，ふち，曲率）をもっているが，◯と + は，大きく異なっている．そのため，脳は普通の 5 の中に逆にした 5 を見つけるよりも，多くの◯の中で + 字を見つける方が簡単である．脳の中にある特徴マップの略図を表した図 7.10 を見てみよう．

しかし，この特徴マップとは何だろうか？　それはどのように定義できるのか？　定義するための 1 つの方法は，フーリエ変換に戻ることである．図 7.10 の例の画像のフーリエマップを取得するとどうなるだろうか？　時間依存信号をフーリエ変換すると，信号に含まれる周波数が得られる．画像は時間

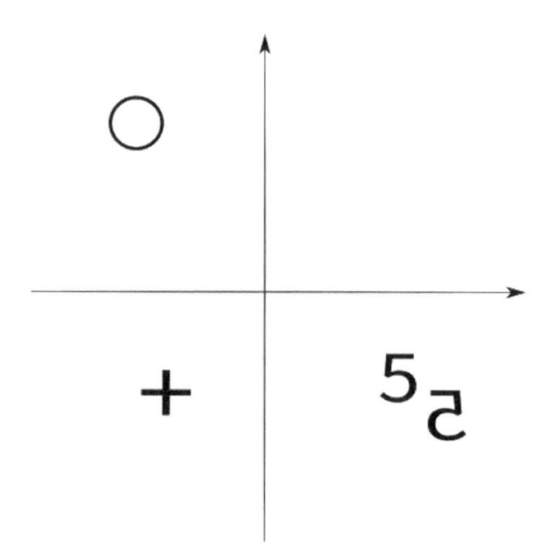

Fig. 7.10 A representation of the feature space. Objects with similar characteristics are located very close to each other, so they are difficult to find in a visual search task. Instead, objects with very different features, like the circle and the plus, are placed in distant positions, so they are easy to distinguish.

But what is this feature map? How can we define it? One way to do that is to go back to our Fourier transform. What happens if we take the Fourier map of the images of the example of figure 7.10? Fourier transformation of time dependent signals results in the frequencies contained in the signal. A picture does not change in time, nevertheless it is possible to calculate a Fourier transformation of it with a two dimensional integral

$$F(\bar{k}) = \iint_A f(x,y)e^{-ik_x x}e^{-ik_y y}dxdy \tag{7.3}$$

What is $\bar{k}$ here? It is the equivalent of ω in the time domain Fourier transformation we discussed previously. In that case $\omega = 2\pi f$ where f was the frequency measured in Hertz. Here $\bar{k}$ is a vector of components $(2\pi k_x, 2\pi k_y)$ where k_x and k_y are the *spatial frequencies* measured in 1/distance (reciprocal of distance). Now, if we have an image composed of N by N pixels i, j, the Fourier transformation will be calculated as follows:

$$F(k,p) = \Sigma_i \Sigma_j f(i,j)e^{-i2\pi\frac{ki}{N}}e^{-i2\pi\frac{pj}{N}} \tag{7.4}$$

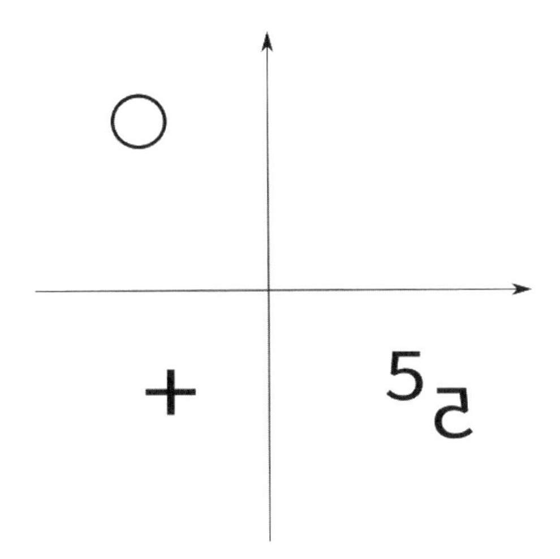

図 7.10 特徴空間の表現．類似した特性をもつオブジェクトは互いに近い位置にあるため，ビジュアル検索タスクで見つけるのは困難である．代わりに，円やプラスのようにはっきりと異なる特徴をもつオブジェクトは，離れた位置に配置されており，そのため区別が容易である．

的に変化しないが，二次元積分でフーリエ変換を計算することは可能である．

$$F(\bar{k}) = \iint_A f(x,y)e^{-ik_x x}e^{-ik_y y}dxdy \tag{7.3}$$

この $\bar{k}$ は何だろうか？ これは，前述した時間領域フーリエ変換における ω と等価である．その場合，$\omega = 2\pi f$ であり，f[Hz] は周波数である．また，$\bar{k}$ は成分 $(2\pi k_x, 2\pi k_y)$ のベクトルである．ここで k_x と k_y は 距離の逆数で測定された**空間周波数** である．ここに $N \times N$ で構成された画像 (i,j) があるとすると，フーリエ変換は次のように計算される．

$$F(k,p) = \Sigma_i \Sigma_j f(i,j)e^{-i2\pi\frac{ki}{N}}e^{-i2\pi\frac{pj}{N}} \tag{7.4}$$

変換の結果は，ピクセルの強度（グレースケール値）が元のイメージの対応する空間周波数を表す別のイメージになる[4]．それでは，このフーリエ変換を前の視覚探索問題に当てはめてみよう．結果を図 7.11 に示す．図の (a) に簡単な「◯の中に + を見つける」という問題，(d) に難しい「5 の中から反

4) 同様に，時間依存信号のフーリエ変換は，各値がその周波数振幅に対応する別の信号である．

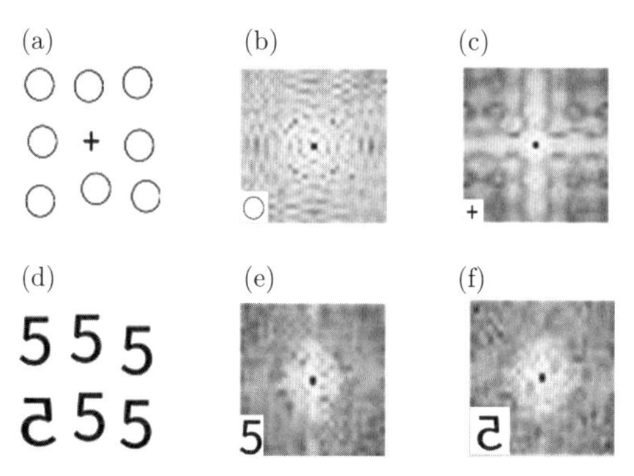

Fig. 7.11 The visual search of different targets and its Fourier transformation. (a) a plus in the middle of many circles and an (d) inverted five among normal fives. The Fourier transformations of circles and plus is shown in the (b) and (c), whereas those of five and inverted five in panels (e) and (f).

Notice that the result of the transformation is now another image where the intensity (the gray-scale value) of the pixel represents the corresponding spatial frequency of the original image[3)]. Now, let's apply this Fourier transformation to the previous visual search problems. The results are shown in figure 7.11. In (a), the easy one: to find a cross among circles, in (d), a more difficult one: find an inverted 5 among normal fives. The Fourier transformation of these characters is shown in the panels on the right side of each problem. You notice how different the transformations are in the case of circle and plus, easy to distinguish. Instead those corresponding to the five and the inverted five appears almost identical, very hard to tell the difference. There is not a direct, proven relation between search efficiency to similarity in Fourier space[Rosenholtz, 2015, Ehinger and Rosenholtz, 2016], however this example suggests that a reason by which the brain have difficulties to solve the latter problem could be this similarity between Fourier transformations.

3) In the same fashion, the Fourier transformation of a time dependent signal is another signal in which each value corresponds to frequencies amplitudes of it.

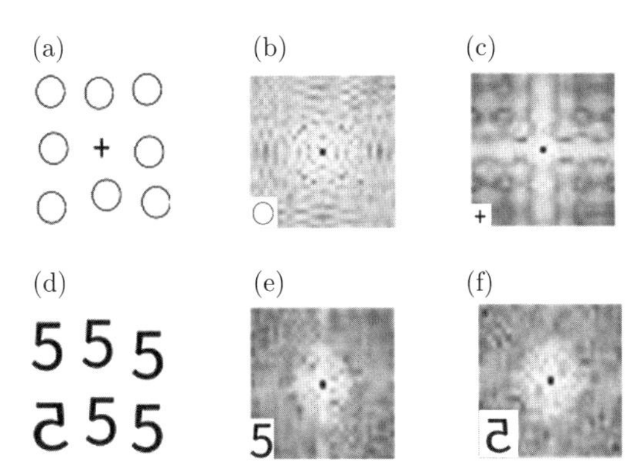

図 7.11　知覚の問題とフーリエ変換後の像．(a) ◯のなかから + を選ぶという問題，(d) 普通の 5 の中に反転された 5 がある問題である．◯と + のフーリエ変換像は (b) と (c) に示す．5 と反転された 5 のフーリエ変換像は (e) と (f) に示す．

転した 5 を見つける」という問題を示している．これらの記号のフーリエ変換像は，◯と + の場合，フーリエ変換した後の図でも簡単に区別がつく．一方，5 と反転された 5 に対応するフーリエ変換像はほとんど同じで，違いを見分けるのが難しい．これらはフーリエ空間での均似性の探索が可能であるという直接の証拠にはならないが，5 の実験では，フーリエ変換後も違いを見分けるのがむずかしいことを示している．[Ehinger and Rosenholtz, 2016, Rosenholtz, 2015].

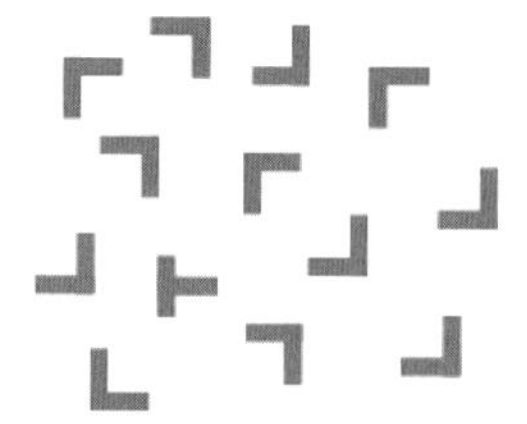

Fig. 7.12 An example of spatial configuration search: find the "T" shape. These types of searches are the most demanding, less efficient. The subject does not have to indicate the exact position of the target, but only clicks a key if he sees the target somewhere and clicks another key if he/she doesn't. Clearly, when the target is absent, subject's vision is forced to scan all the items to confirm that the target is not present and this takes longer.

7.4 Search efficiency: seconds per item

Can we measure how good the visual system is in a search? Yes, we can use the *search efficiency* parameter. This value is obtained directly by experiments of visual search and it is measured generally in milliseconds per item. For example let's suppose that we ask a participant to find an inverted "5" character among a panel of N normal ones, like in figure 7.7. Depending on the number of items N, we will have an average reaction time RT. We can plot the RT in function of the number of items, the dependence is linear and the inclination of the line is called search efficiency. The search efficiency depends on the kind of search we are doing. In the case of figure 7.9 the curve has almost flat inclination (high efficiency) whereas in the other case it is even higher. Psychologists have classified the visual search tasks into *features search* (easy, with about zero milliseconds per item efficiency), *conjunction search* (about 5–15 ms per item) and *spatial configuration search* (the most difficult, 20–30 ms per item, see figure 7.12 for example).

In addition to what we said above, please consider what has been discovered about the visual system:

- it is sensitive to time derivative: blinking objects, moving objects attract the attention.

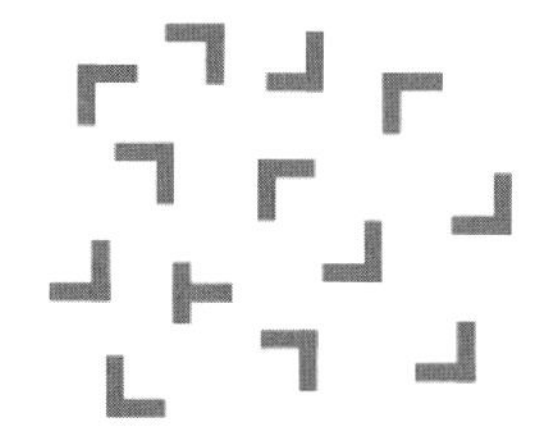

図 7.12　空間構成探索の例："T" 字形を探してみよう．このタイプの検索は最も要求が厳しく効率が悪い．被験者はターゲットの正確な位置を示す必要はないが，ターゲットがどこかにある場合はキーをクリック，ターゲットが見えない場合には別のキーをクリックする．明らかに，ターゲットが存在しない場合，被験者の視覚はすべてのアイテムをスキャンしてターゲットが存在しないことを確認する必要があり，これにはさらに時間がかかる．

7.4　探索効率：1 アイテムあたりの探索時間

視覚システムがどの程度探索に適しているかを測定できるか？　私たちは**探索効率パラメータ**を使うことができる．この値は視覚的な探索の実験によって直接得られ，それは一般的に 1 アイテムあたりミリ秒で測定される．例えば，図 7.7 のように，N 個の通常のパネルから反転した「5」の文字を見つけるように被験者に依頼する．アイテムの数 N に応じて，平均反応時間 (Reaction Time) RT が得られる．RT を項目数の関数でプロットすることができる．依存性は線形であり，傾きは探索効率と呼ばれる．この探索効率は，探索の種類によって異なる．図 7.9 の場合，曲線はほぼ平坦な傾斜，つまり高効率を有する一方，他の場合ではより高くなる．心理学者は，視覚探索タスクを**特徴探索**（最も簡単だと，項目あたり約 0 ミリ秒かかる），**接続探索**（1 項目あたり約 5〜15 ミリ秒かかる）および**空間構成探索**（図 7.12 のように，最も困難で，1 項目ごとに約 20〜30 ミリ秒かかる）に分類した．

これまでの内容を下記にまとめたので，視覚システムの科学でわかっていることは何であるか考えてみよう．

- 時間変化に対する反応：点滅するものや，動くものへの注意．
- 空間変化に対する反応：明確な境界線はよく気がつくが，なめらかなグラデーションは気がつきにくい．
- 時間依存の周波数と空間依存の周波数：時間（周波数）と空間（波長）に

- it is sensitive to space derivative: we notice edges, we do not notice smooth variations of colors.
- it is sensitive to time-domain frequencies *and* space-domain frequencies. This means that the Fourier transformation in time (frequency in Hz) and in space (wavelength) are important factors in the perception of a scene.

The visual system is very complex and its functioning is still unknown in its fine details. However, in this chapter we gave a simple description of the basic fundamental principles by which it operates. Now please consider these questions:

Think about these questions

- What is the Gestalt theory?
- Can you describe the *proximity principle*?
- What does the principle of *emergence* have to do with that of *closure*?
- What is the principle of *similarity*?
- What is the *Fourier* transformation? Why is it important to perception?
- Can you describe a *visual search* problem?
- What is the *feature map*?
- Why is it easier to spot a cross among circles, than a "5" among inverted "5"s?

対するフーリエ変換は，情景認知の重要な要因である．

視覚システムは，とても複雑で，そのはたらきの詳細はまだ解明されていない．しかし，本章でのいくつかの作業によって，その基礎的な原則の表現を得た．ここで次の問題を考えてみよう！

考えてみよう

- ゲシュタルト理論とはどのようなものか？
- 近接の原則とはどのようなものか？
- 出現の原則と閉鎖の原則には，どのような関係があるか？
- 類似の原則とはどのようなものか？
- フーリエ変換とはどのようなものか？　また，知覚における重要な点はどのようなものか？
- 視覚探索の問題例はどのようなものがあるか？
- 特徴マップとはどのようなものか？
- ○の中から×を見つけ出すことが，反転した5の中から5を見つけ出すことより簡単なのはなぜか？

Chapter **8**

A new kind of neuron

As seen in the previous chapter 2, artificial neural networks (ANNs) have limited abilities: some are subjects to *supervised* learning process, some other are good to solve certain types of problems, others can solve different ones. Nevertheless, real brain intelligence is much better, biological intelligence is *general*. We are not limited to specific problems and we do need particular supervised training. The artificial neural networks are successful and astonishingly good, but still do not reproduce the generality of human intelligence. For this reason, researchers are looking for new solutions.

In 2004 Jeff Hawkins[Hawkins, 2004] devised a new paradigm for neural networks that we are going to briefly introduce in this chapter. To grasp the basics of this new idea we have to touch the following important concepts.

8.1 Layers and columns

Neurons are organized in *layers* and *columns*[Hawkins and Ahmad, 2016], see figure 8.1. Layers are sets of neurons that are communicating to each other in an intricate network of connections. The majority of the neurons in a layer are connected to other neurons in the same layer and their connections can extend to long distances. Columns, instead, are set of neurons connected in short distances and in a feed-forward manner. The main characteristics of columns is that all the neurons in a column share the same input connections. Cells in a column are often connected laterally to other columns of the same layer. We can schematize layers and columns as in the sketch of figure 8.1.

第 8 章

最新のニューラルネットワーク

ニューラルネットワークの進歩に伴って，まったく新しい発想のニューラルネットワークの必要性が高まっている．ニューラルネットワークは，人工知能の分野での成功を収めたが，未だに実際の脳には及ばない．前章までのニューラルネットワークでは，問題ごとに学習過程を設定する必要があり，ある問題に対しては解決可能だが，他の問題では解決不可能という場合があった．しかし，私たちの知能はとても汎用性があり，特定の問題に制限されることなく，個々に適した学習を行うことができる．ニューラルネットワークの研究が進み，飛躍的に向上しているが，人間の知能の高い汎用性を生み出すことはできていない．2004 年，ジェフ・ホーキンスによってニューラルネットワークの新たな指針が示された [Hawkins, 2004]．この章では，それらを簡潔に解説する．この新しい考え方の本質を捉えるために，いくつかの重要なコンセプトを紹介しよう．

8.1 層とカラム

ニューロンは図 8.1 に示すように，**層とカラム**によって構成されている [Hawkins and Ahmad, 2016]．層には，情報伝達のために互いに結合された複数のニューロンが配置されている．層内の大半のニューロンは，同じ層内の他のニューロンと結合する．そして，その結合は，空間的に長い距離でも成立する．一方，カラムは，配置されたニューロンの中で短い距離でフィードフォワード的に結合する．そして，その結合は，フィードフォリード的である．カラムの最大の特徴は，1 つのカラム内のすべてのニューロンは，同じ入力を共有している点である．あるカラムのニューロンは，同じ層の他のカラムと結合されることもある．層とカラムの模式図を図 8.1 に示す．

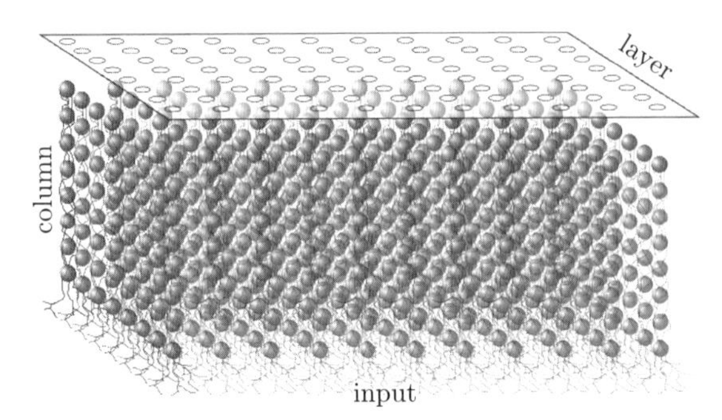

Fig. 8.1 The structure of layers and columns. A *column* is a set of neurons that share the same input pattern and are feed-forwardly connected. The *layer* can be imagined like the ensemble of all the column's outputs. On the bottom the dendrites are represented by branching filaments that connect to the input (each column receives a different input pattern). This system of neurons can store an enormous amount of data.

8.2 Neural inhibition: the first to spike wins!

When a neuron becomes active, it inhibits all the neighbor ones. This is a very important and fundamental mechanism in the brain. It reduces the number of active neurons to a limited value. In the human cortex, the ratio of active neurons to not-active neurons is very low, only about 3–4%.

This ratio can be called also with the term *sparsity*. So in a column, the first neuron to spike wins and inhibits the other neurons in the column to do the same. A similar phenomenon happens also between columns. The first column to become active (a column becomes active when at least one of its neurons is active) induces neighbor columns to be inert. This mechanism maintains constant sparsity in layers.

8.3 A new neuronal input paradigm: proximity and periphery coincidence

In this cortical model, there are two different kinds of inputs (see figure 8.2).

As we know, signals arrive to the neuron through synapses that are lo-

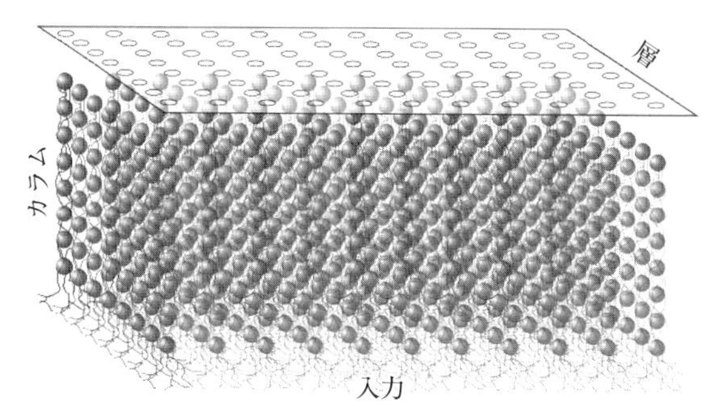

図 8.1　層とカラムの構造を示した．カラムに配置されたニューロンは，入力パターンを共有し，フィードフォワード的に結合している．層は，カラムの出力の総和とイメージすることができる．樹上突起の末端は，入力に接続される細い根のように表される（それぞれのカラムは，異なる入力パターンを受け取る）．つまり，このニューロンのシステムは，大量のデータを蓄積できる．

8.2　ニューロンの抑制，初期発火の優位性

あるニューロンが活動電位に達したとき，他の近距離のニューロンは発火を抑制される．これは脳において非常に重要で基本的なメカニズムである．また，同時に発火するニューロンを制限された数に減らすはたらきがある．人間の大脳皮質全体でさえ，発火しているニューロンの数は，約 3〜4% と非常に少ない割合である．

この割合は，位相幾何学の用語でスパーシティー（疎行列）と呼ばれている．つまり，カラム内で最初に発火したニューロンが他のニューロンを抑制するはたらきに相当する．同様の現象は，カラム間でも起こる．最初のカラムが活性化すると，近隣のカラムの不活性化を引き起こす（カラムが活性化すると，少なくとも 1 つのニューロンが活性化する）．このメカニズムによって，層内のスパーシティーは一定に保たれる．

8.3　ニューラルネットワークにおける入力の新しい枠組み：近接性や同時性について

いま議論している新しい皮質のモデルには，図 8.2 に示すように，入力に 2 つの異なる種類がある．

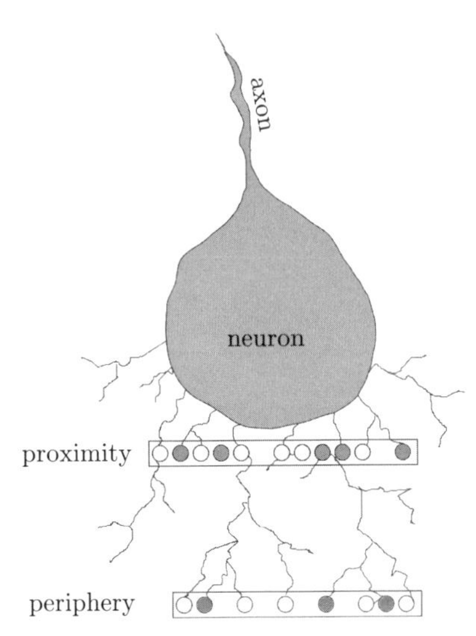

Fig. 8.2 According to biological experiments, the neuron responds differently to input in the *proximity* of the cell body, than to that in the *periphery*. The thin filaments represent dendrites. Along them active synapses are represented by black circles and the not active by white ones.

cated along filaments called dendrites. Those coming from the *proximity* of the cell body provoke a signal only if there is synchronization between synapses. This can be modeled as a time *coincidence* input. With the term coincidence we mean temporal synchronization. In other words, if along the dendrite different synapses receive spikes almost simultaneously, the neuron will spike. Of course, too few neurons will not be enough, there should be a minimum number of simultaneous spikes.

A different kind of neural input comes from synapses in the *periphery*. This means regions on the dendrites far away from the neuron body. Because of the distance, the coincidence does not induce the neuron to spike, but it only depolarizes it. This puts the neuron in a state of readiness to spike, the so called *prediction* state. When the neuron is in such a state and a proximity signal comes, it spikes *sooner* than normal.

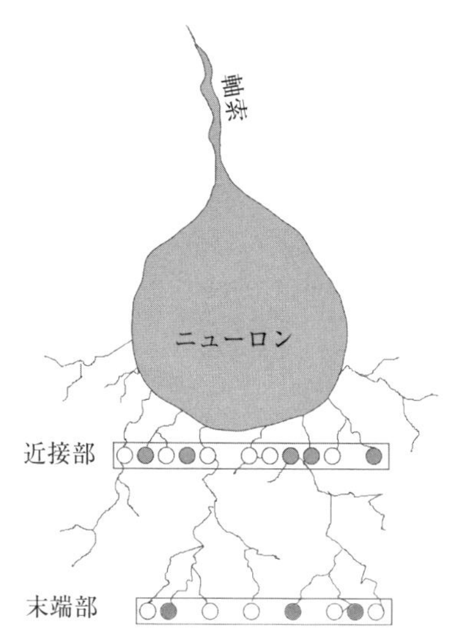

図 8.2 生物学的な実験によると，ニューロンは細胞体に近い入力と，末端の入力では異なった反応を示す．細い枝のようなものは，樹上突起である．活性なシナプスは黒丸で不活性なシナプスは白丸で示した．

すでに学んだように，信号はシナプスの樹上突起と呼ばれる細い糸を通じてニューロンに伝達される．細胞体との近さに起因して，シナプス間の同期性に関する信号を誘発する．これは，同時性の高い入力とモデル化ができる．同時性とは，一時的な同期を意味する．言い換えると，もし異なるシナプスが樹上突起を通じて，ほぼ同時に発火の信号を受け取ったとすると，そのニューロンも発火を誘発する．もちろん，同時に発火する最小の数があるので，ニューロンが少なすぎる場合には起こりえない．

また，異なる種類のニューロンへの入力は，シナプスの末端から伝達される．これは，ニューロンの本体から遠い樹上突起の領域を意味する．その距離によって，ニューロンの発火の同時性は妨げられるが，ニューロンを脱分極状態にする．予測状態と呼ばれる発火の準備ができている状態になる．ニューロンが予測状態にあり，近距離で信号が伝達された場合，普段よりも早期に発火する．

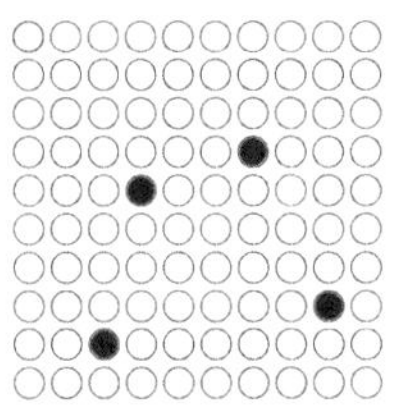

Fig. 8.3 The sketch of a neuronal layer with 100 columns. The circle represents the activity of a column. A column is active (represented in black) if at least one of its neurons is spiking. Because of inhibition, the active neurons are always about 3-4% of the total (the *sparsity* in mammal brain is about 2-5%). A layer like this can represent millions of different patterns.

8.4 Sparse representation capacity

Imagine a layer as an ensemble of neuronal columns. A column is active when at least one of the neurons in the column is active, so we can depict a layer as a set of active or not-active elements (see figure 8.3). There is a strong experimental evidence that shows that layers activity holds *semantic* positional information. This means that if a particular neuron in a layer is active, that neuron holds a specific meaning. The neuron can be activated when a certain action is taken, or when a certain object is seen. Thus, a pattern of active neurons, represents a set of different meaningful actions or perceptions. Altogether, this set may represent a complex action, a detailed perception or an abstract higher level concept.

Now, we can ask ourselves: how many patterns can a layer store?

Suppose we have a layer of only 5 columns: each column can be only active or not active, so the total combinations are $2^5 = 32$. This number grows exponentially, if we have a layer of 10 by 10 columns, the combinations rise to 2^{100}, this is a number with 30 zeros(!). So the capacity of a layer seems really enormous. However, we know that inhibition is at work in mammal brains, so we suppose that only 4% of the columns are active at any time. How many patterns combination are possible? This is equivalent to the *combinations* of w objects in n possible positions. In combinatorial

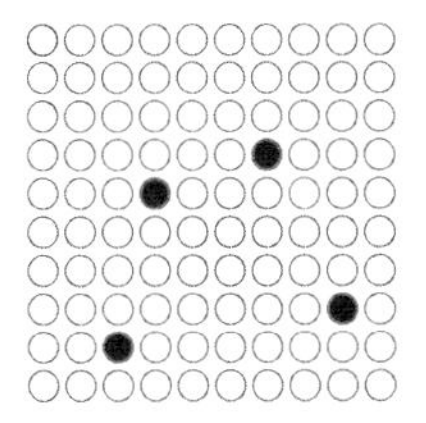

図 8.3　図は 100 個のカラムからなるニューロンの層の模式図．◯はカラムが活性かどうかを示している．黒丸では，カラムの中で少なくとも 1 つのニューロンが発火していることを示している．抑制のルールにより活性なニューロンは全体の 3〜4% 程度である（実際の脳内でのスパーシティーも 2〜5% 程度である）．各層には，このような数百万通りの異なるパターンが存在する．

8.4　スパーシティーによる表現と容量

ここで，カラムが整列した層について考えてみよう．カラムは，そのカラム内の少なくとも 1 つのニューロンが活性であれば，活性であると言える．つまり，活性または不活性な要素を配置することで各層を描画できる（図 8.3）．各層の活性は意味の配置情報を保持しているとする，重要な実験的証拠が存在する．つまり，ある層の特定のニューロンが活性な場合，そのニューロンは特定の意味をもつということである．ある特定の行動や特定のものを見たとき，ニューロンは活性になることが許される．つまり，活性なニューロンのパターンは，異なる意味の行動や知覚の配置を表現している．総合すると，この配置で複雑な行動や詳細な知覚や高次元の抽象的な概念も表現可能かもしれない．

ここで，層にはどのくらいのパターンが蓄積できるのかを自分自身に問いかけてみたい．

ここに 5 つのカラムからなる層を仮定しよう．各々のカラムの状態は活性か不活性の 2 状態があるため，組み合わせの総数は $2^5 = 32$ になる．この数字は指数関数的に増加する．もし 10×10 のカラムからなる層であれば，組み合わせの総数は 2^{100} に跳ね上がる．これは 31 桁の数字である．各層の容量は，本当に膨大に思える．しかしながら，実際の脳では抑制のはたらきがあり，いかなる時でも，カラムの活性は 4% にすぎない．一体どのくらいの数の組み合わせが存在するのだろうか？　これは，w 個のものを n マスの盤面に並べる組み合わせと同等である．統計学的に組み合わせを考えれば，組み合

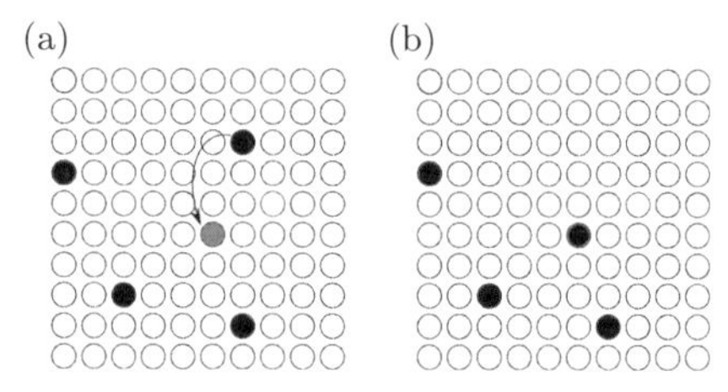

Fig. 8.4 What happens if we change one of the active elements in this 100 element layer? The brain can easily recognize pattern (b) as (a) even moving away one of its active elements (one out of four corresponds to an error of 25%!). This is because the probability that the other three exactly match the pattern by chance is extremely small. Sparsity give robustness to the system. If activity is not sparse (for example 50% of the neurons were active), two different patterns would have a lot of points in common and could cause confusion, instead the probability of mismatch becomes smaller with sparse activation).

statistics it is calculated using the symbol ${}_nC_w$ that is:

$$ {}_nC_w = \binom{n}{w} = \frac{n!}{w!(n-w)!} \tag{8.1} $$

in our case of $n = 100$ and $w = 4$ this gives ${}_{100}C_4 \approx 4 * 10^6$. With a layer of just a hundred neurons, millions of pattern can be represented. Just imagine if the layer is bigger than that. Try to calculate some examples by yourself!

8.5 Robustness to noise

We understood in the previous section that layer of active or inactive neurons can represent an enormous quantity of patterns. But how reliable are these representations? Let's test the robustness to noise. We add a certain amount of noise to our pattern with $n = 100$ neurons and $w = 4$ active ones in the example above. Let's call this pattern Θ. We now introduce 24% of noise, that is we change one of the 4 active neurons. We now have a new pattern $\bar{\Theta}$, of which three of the pixels coincide with Θ but one is in a different position (see figure 8.4).

Can the brain still recognize that the pattern $\bar{\Theta}$ is originated from Θ? Or, a better question is: what is the probability that two patterns of four

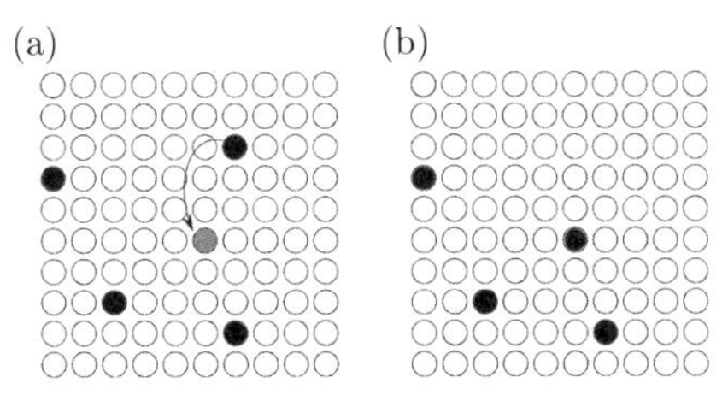

図 8.4　100 個の要素からなる層のうち，活性な要素の 1 つを変更すると何が起こるだろうか？　脳はパターン (b) が 1 つの活性な要素の移動を除いてはパターン (a) と一致していることを簡単に認識できる（4 つのうち，1 つが異なるということは，そのエラーは 25% になるにも関わらず）．これは，他の 3 つのパターンが一致する可能性が非常に小さいことに起因する．スパーシティーは，システムのロバスト性を担っている．もし，活性化にスパーシティーがなければ，2 つの異なるパターンは，多くの共通部をもつことで混乱を招くことになる（例えば 50% のニューロンが活性であった場合など）．言い換えると，活性のスパーシティーによって，不一致の可能性を低減できる．

わせ ${}_nC_w$ を用いて次のように表される．

$$ {}_nC_w = \binom{n}{w} = \frac{n!}{w!(n-w)!} \tag{8.1} $$

$n = 100$ で $w = 4$ の場合，その組み合わせは ${}_{100}C_4 \approx 4 \times 10^6$ で与えられる．100 ニューロンの層では，数百万通りのパターンを表現できる．さらにもっと大きい層も考えられる．いくつかの具体的な例で計算してみよう．

8.5　ノイズに対するロバスト性

私たちは，すでに活性なニューロンや不活性なニューロンの層によって膨大なパターンを表現できることを理解した．しかし，それらの表現はどの程度信頼できるのだろうか？　ここで例として，$w = 4$ 個が活性な $n = 100$ 個のニューロンからなるパターン（表現）にノイズを加えてみよう．今これをパターン Θ と呼ぶ．ここに 24% のノイズを付加する．すなわち 4 つの活性なニューロンのうち，1 つを変更するということである．これにより，新たなパターン $\bar{\Theta}$ を得る．3 つのピクセルは Θ に一致するが，1 つは異なる場所に位置する（図 8.4）．

脳は，パターン $\bar{\Theta}$ がパターン Θ からの派生であると認識できるのだろうか．これは非常に面白い問いである．100 個のニューロンのうち，4 つの活性

active neurons out of 100, have three of them exactly coincident? You can attempt a very rough calculation by yourself: the probability that one neuron is active exactly in one position is $p = 1/100$. The probability of having a match of three of them is $p \approx 10^{-6}$, one out of a million! The calculation is simplistic (we did not take overlap into account), but it shows that the layer is extremely robust against noise, even with only 100 neurons. You can find exact calculations of these probabilities in the work of Ahmad S.[Ahmad and Scheinkman, 2019]. There you can realize that layers of few thousand neurons can encode an enormous amount of information and that sparsity gives a great help to avoid mismatches between patterns. In fact, the encoding is incredibly robust to noise, it is possible to calculate exactly the robustness to false positives (mismatch between patterns). For example adding up to 30% of noise to a layer of about 2000 neurons with sparsity of 2%, the probability of mismatch is of the order of 10^{-50}, a number so small that is practically zero (see for example M. Taylor[Taylor, 2015]).

8.6 Prediction

The human brain behavior suggests that it functions by predicting sequences (see chapter 2). Suppose that each neuron in a column is connected not only forward to the adjacent neuron, but also laterally to other columns. Because of the distance, the connection will be peripheral, so it will not directly induce neurons to spike, but it will put them in depolarized state. This means that those neurons will spike sooner than other neurons with the same input.

Suppose there is a sequence of input patterns coming in the network. Let's call these "A", "B" and "C". Suppose also that our network has only 10 columns. Input "A" will activate for example columns 2, 5 and 7 (see figure 8.5 for a sketch of this example).

Let's imagine that after pattern "A", pattern "B" follows. One of the neurons in each columns of input "B" (1, 3 and 9 in our example) will

なニューロンがある2つのパターンにおいて，それらの3つが完全に一致する確率はどのくらいか？　大雑把な計算で確認してみよう．ある位置に1つの活性なニューロンが一致する確率は，$p = 1/100$ であり，3つの活性なニューロンが一致する確率は，$p \approx 10^{-6}$ である．なんと百万分の一になる．たった100個のニューロンを用い，重複を数えない簡易的な計算であるが，各層のノイズに対するロバスト性が示された（たった100個のニューロンであっても）．アハマドの論文からも確率計算が正しいことが確認できる [Ahmad and Scheinkman, 2019]．ここで，数千個のニューロンからなる層は，大量の情報を読み込むために必要であり，またそのスパーシティーはパターンの不一致を避けるために非常に有効である．実際，読み込みには多くのノイズを含んでいるがロバスト性は保たれている．また，不一致などの誤りに対するロバスト性は，計算し，数値化することが可能である．例えば，ノイズを30% 含んだ，約2000ニューロンからなるスパーシティーが2% の層の不一致の確率は 10^{-50} のオーダーである．これは実用上は0と見なすことができる（[Taylor, 2015] を参照）．

8.6　予測

第2章に述べたように，人の脳のふるまいには，順序を予測するはたらきがある．あるカラムのニューロンが，それぞれ隣り合うニューロンと結合するだけでなく，他のカラムとも結合していると仮定しよう．末端での結合はカラムとカラムの距離が遠くなるため，ニューロンの発火が直接的に誘発されることはないが，活性化が脱分極の状態になる．これは，それらのニューロンが，同じ入力の他のニューロンよりも早く発火することを意味する．

ここで，あるネットワークに順序の入力パターンが伝達されたと仮定する．それらのシーケンスを “A”, “B”, “C” と呼ぶ．またネットワークは10個のカラムを仮定する．入力 “A” は，2, 5, 7のカラムを活性化する（例として図8.5に模式図を示した）．

“A” のパターンの後に，“B” の入力がある場合を考えよう．パターン “B”（1, 3, 9の入力）の入力によって各カラムのニューロンの1つはパターン “A” により活性化されたニューロンによって予測状態にある．パターン “B” の入

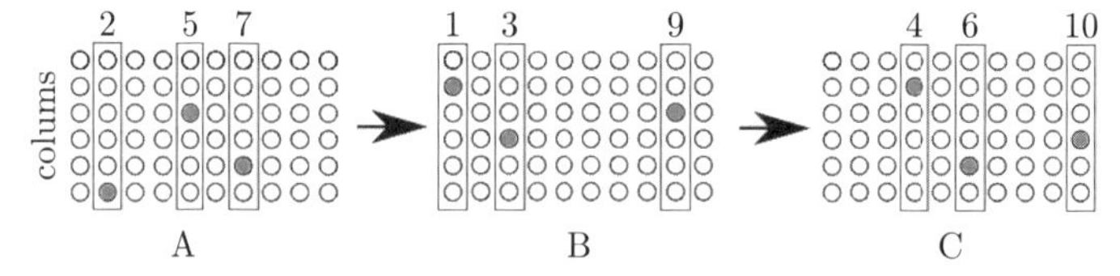

Fig. 8.5 A sketch of sequence learning in HTM. Different input patterns activate columns of neurons. We suppose that only one neuron is active in each column. The active neuron is connected to other neurons of other columns. As a result, those neurons become depolarized. When the column is activated, that neuron will fire first and inhibit all the others in that column. In this way, the same pattern of active columns can be represented by many patterns of neurons within the columns. These different patterns are called *contexts*. The possible contexts for the same columns pattern is the number of neurons in each column to the power of the number of active columns. In this 10 columns example there are more than hundred combinations. This number gets soon very big for realistic cases. The brain can memorize the same pattern in a huge number of different *contexts*!

be put in predictive state by the neurons activated by pattern "A". Then when pattern "B" arrives, that neuron will be the first to spike. Because of inhibition, the other neurons in that column will not spike. That active neuron will put other neurons of other columns in predictive state. Similarly, when pattern "C" arrives (signal in columns 4, 6 and 10) only those neurons in predictive state will be activated. If the sequence "A", "B" and "C" is repeated often, Hebbian learning will make the synaptic strengths of those connection stronger and stronger. So, the sequence "A"→"B"→"C" will be learned. Every time "A" comes, the neurons in the columns for "B" will be in predictive state. The pattern of active neurons of input "B" will be a unique representation of pattern "B" when it comes after object "A". We can say that *pattern "B" is in the context of pattern "A"*. By the same mechanism, we can also say that pattern "C" is in the context of "A"→"B".

What happens if we have another pattern "X" instead of "A"? The neurons in the predictive state will be different, so when "B" will be in input, other neurons in the columns 1, 3 and 9 will be active! So, the active columns for "B" will be the same, but the active neurons within the columns will be different. Since "B" in context of "A" is different from "B" in context of

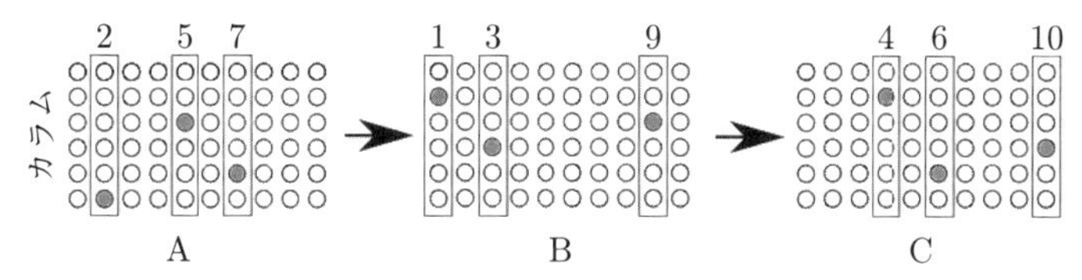

図 8.5 HTM における学習の順序の例．異なる入力パターンは，カラムのニューロンを活性化する．各カラムで 1 つのニューロンのみ活性化する．活性化したニューロンは，他のカラムの他のニューロンに結合される．その結果，それらのニューロンは予測状態になる．カラムが活性化されたとき，最初の発火により，そのカラム内の他のニューロンは抑制される．この方法により，活性化されたカラムは，カラム内の異なるニューロンで複数のパターンとして表される．それらの異なるパターンは文脈（前後関係）と呼ばれる．同じカラムのパターンによる文脈の通りの数は，各カラムのニューロンの数に活性化されたカラムの数を乗じた数字になる．この図の例では，100 以上の組み合わせが可能である．6 の 3 乗である．この数字は現実には，とてつもなく巨大な数字になる．脳はいくつもの似たような文脈の異なるパターンを記憶することができる．

力が到達したとき，ニューロンは最初の発火に抑制され，同じカラムの他のニューロンの発火は遅れるだろう．活性化したニューロンは，他のカラムの他のニューロンを予測状態にする．パターン “C”（4, 6, 10 のカラムの信号）が到達したとき，予測状態にあるニューロンのみ活性化される．もし，“A”, “B”, “C” の順で繰り返し信号が到達すれば，ヘッブの学習則によってシナプス同士の結合はより強化される．“A”, “B”, “C” の順で繰り返し学習することで，“A” の信号が到達すると，そのカラムのニューロンは “B” の予測状態になる．“A” の信号が到達したあとの入力 “B” の信号は，特定のパターン “B” を表す．今，私たちが言うパターン “B” とは，文脈 “A” としてのパターン “B” である．同様なメカニズムによって，パターン “C” とは，文脈 “A” そして “B” としてのパターン “C” である．

もし，パターン “A” の代わりにパターン “X” の信号が到達したらどうなるだろうか？ パターン “B” が到達した場合とは違う別のニューロンが予測状態になり，カラム 1, 3, 9 の別のニューロンは活性化されるだろう．つまり，活性化されるカラムは同じだが，活性化されるニューロンは異なる．文脈 “A” としてのパターン “B” と文脈 “X” としてのパターン “B” が異なるため，もちろんパターン “C” にも影響がある．文脈 “A” そして “B” としてのパターン “C” と文脈 “X” そして “B” としてのパターン “C” は異なる．当然パタ

"X", also "C" will be influenced. "C" in context of "A"→"B" is different from "C" in context of "X"→"B". Of course, instead of "C" we can have a different column pattern "D". The brain can also predict different outcomes for "B" depending on whether the previous pattern was an "X" or an "A". This is the way our brain memorizes high order sequences! The brain is able to recognize the same pattern of active columns in millions of different ways. Each way represents a different *context* by which that pattern appears. The storage capacity of this system is:

$$sC = (cN)^{aC} \tag{8.2}$$

where cN is the number of neurons per each column and aC is the number of active columns. This value gets immediately huge, in our example will be six to the power of three that is already more than hundred possible patterns, but you can just imagine the storage power of a real brain where neural layers have thousands of even millions of active columns!

8.7 Hierarchical temporal memory (HTM)

The hierarchical temporal memory (HTM) is the name given to an artificial neural networks based on the concepts that we have learned in this chapter. The network is working as a memory system of unbelievable store capacity. The system doesn't learn static data, but learns sequences of dynamically changing streams of data. The data are represented by patterns of active or non-active columns of neurons in a layer. The neurons are activated in a *sparse* manner, this means that just 2–5% of them are active at any moment. Sparsity reduces the capacity, but gives a great advantage in robustness to noise. As we have shown in the previous sections, if we introduce in a layer of only 100 columns about 30% of errors in the data, still we have extremely low probability of mismatch. The system is able to store sequences of patterns very reliably, and each pattern is recognized in the context of preceding data. This is how the brain is able to choose

ーン "C" の代わりにパターン "D" が到達する可能性もある．脳は，パターン "X" かパターン "A" に依存して，パターン "B" は異なる出力を予測する．私たちの脳内では，膨大な長さのシーケンスを記憶している．脳は，活性なカラムが一致しているパターンを，異なる方法でいくつも認識している．どのパターンが現れるかという異なる文脈によって別々に表現される．このシステムにおける蓄積の容量は下記の通りである．

$$sC = (cN)^{aC} \tag{8.2}$$

ここで，cN は各カラムのニューロンの数で，aC は活性なカラムの数である．これは，とても膨大な値になる．この例題で見ても 6 の 3 乗となる．この例題ですでに 100 通りを越えるパターンを蓄積しているが，実際の脳におけるニューロンの層を考えると活性なカラムが何十億にのぼる．

8.7 階層型一時記憶 (HTM)

階層型一時記憶 (hierarchical temporal memory：HTM) は，この章で学んだ概念を基礎にした人工ニューラルネットワークである．ネットワークは膨大な容量の記憶装置としてはたらく．そのシステムは統計的な学習ではなく，データの流れの変化を動的に処理する．データは層内のニューロンが活性または不活性のパターンで表される．その活性化されたニューロンには，スパーシティーの規則が適用され，同時に活性化される割合は 2〜5% にすぎない．スパーシティーは，容量を減少させるが，ノイズに対するロバスト性を保つためには，有利にはたらく．前節で計算したように，100 個のカラムをもつ層に導入するノイズが約 30% あっても，データが不一致になる可能性はごく低い．また，そのシステムはパターンの順序も蓄積可能である．そして，互いのパターンを文脈として予測することで認識可能である．これは，異なる流れの中で起こる状況に対して，脳が何をするべきかを選択する方法を示したものになる．HTM システムは，人工ニューラルネットワークの一種で，CNNs や

what to do next in a situation that happens in different context. The HTM systems are artificial neural networks, but are not yet able to solve classification problems well as CNNs or RNNs systems. However, HTM networks are incredibly robust to noise and they are currently used for anomaly detection and prediction applications. Moreover, HTMs are based on realistic biological processes, so are very promising to the goal of realizing biological plausible artificial brains.

Think about these questions

- What is a *column*?
- What does *inhibition* mean?
- What is *sparsity*?
- What is the difference between proximal synapses and peripheral ones?
- What are the advantages and disadvantages of sparsity?
- How much is the storage capacity of a neuronal layer of 2000 column at 4% sparsity?
- How does the brain predict sequences of patterns?
- What is the *context* in HTM sparse representation?
- What does HTM means?
- What are the most promising applications for HTM?

RNNsのように分類問題に対する解決策の応用はまだ少ない．しかしHTMシステムは，高いロバスト性をもち，すでに異常検知や予測問題に対する応用例がある．また，実際の生物学的な事実と深い関わりに基礎をなしているため，近い将来，きっと生物学的にもっともらしい人工知能の実現を約束してくれるだろう．

考えてみよう

- カラムとはどのようなものか？
- 抑制とは何を意味しているのか？
- スパーシティーとはどのようなものか？
- 細胞体に近い入力と末端の入力にはどのような違いがあるか？
- スパーシティーが，有利にはたらく点，不利にはたらく点はどのようなものか？
- 4% のスパーシティーをもつ 2000 のカラムからなる層の蓄積容量はどの程度か？
- 私たちの脳は，どのような方法でパターンのシーケンスを予測するのか？
- HTM における文脈とはどのように表されるのか？
- HTM はどのような意味をもつのか？
- HTM の応用例として期待されることは何か？

Bibliography／参考文献

[Ahmad and Scheinkman, 2019] Ahmad, S. and Scheinkman, L. (2019). How can we be so dense? The benefits of using highly sparse representations. In *ICML2019 Workshop on Uncertainty and Robustness in Deep Learning.*

[Ashmore and Geleoc, 1999] Ashmore, J. and Geleoc, G. S. (1999). Cochlear function: Hearing in the fast lane. *Current Biology*, 9(15):R572–R574.

[Brown *et al.*, 2004] Brown, E., Kass, R., and Mitra, P. (2004). Multiple neural spike train data analysis: state-of-the-art and future challenges. *Nature Neuroscience*, 7(5):456–461.

[Castelfranchi, 2013] Castelfranchi, C. (2013). Alan Turing's "Computing Machinery and Intelligence". *TOPOI-An International Review of Philosophy*, 32(2):293–299.

[Chollet *et al.*, 2015] Chollet, F. *et al.* (2015). Keras. `https://keras.io`.

[Edalat, 2019] Edalat, A. (2019). Hopfield networks. `https://www.doc.ic.ac.uk/~ae/papers/Hopfield-networks-15.pdf`. [Online; accessed 11-Feb-2019].

[Ehinger and Rosenholtz, 2016] Ehinger, K. A. and Rosenholtz, R. (2016). A general account of peripheral encoding also predicts scene perception performance. *Journal of Vision*, 16(2):13, 1–19.

[Hawkins, 2004] Hawkins, J. (2004). *On intelligence.* Henry Holt & C., New York.

[Hawkins and Ahmad, 2016] Hawkins, J. and Ahmad, S. (2016). Why neurons have thousands of synapses, a theory of sequence memory in neocortex. *Frontiers in Neural Circuits*, 10:23.

[Hodgkin and Huxley, 1952] Hodgkin, A. L. and Huxley, A. F. (1952). A quantitative description of membrane current and its application to conduction and excitation in nerve. *The Journal of physiology*, 117:500–544.

[Hong *et al.*, 2012] Hong, S., Ratté, S., Prescott, S. A., and De Schutter, E. (2012). Single neuron firing properties impact correlation-based population coding. *Journal of Neuroscience*, 32(4):1413–1428.

[Hopfield, 1982] Hopfield, J. (1982). Neural Networks and physical systems with emergent collective computational abilities. *Proceeding of the National Academy of Science*, 79:2554-2558.

[Izhikevich, 2000] Izhikevich, E. (2000). Neural excitability, spiking and bursting. *International Journal of Bifurcation and Chaos*, 10(6):1171-1266.

[Izhikevich, 2003] Izhikevich, E. (2003). Simple model of spiking neurons. *IEEE Transactions on Neural Networks*, 14(6):1569-1572.

[Izhikevich, 2010] Izhikevich, E. M. (2010). *Dynamical systems in neuroscience.* The MIT press.

[Kaas, 2011] Kaas, J. H. (2011). Neocortex in early mammals and its subsequent variations. *New perspectives on neurobehavioral evolution* (ed. Johnson, J.I. *et al.*), volume 1225 of *Annals of the New York Academy of Sciences*, pp.28-36. Neuroscience Associates Incorporated; Michigan State University, Department of Radiolgy. Conference on New Studies of Neurobehavioral Evolution, Armed Forces Institute of Pathology, Washington, DC, JUN 25-28, 2010.

[Kaňok, 2018] Kaňok, T. (2018). Appled demonstration of convolution in one dimensions. https://www.fit.vutbr.cz/study/courses/ISS/public/demos/conv/. Faculty of Information Technology, University of Technology, Brno, CZ.

[Karp, 2008] Karp, G. (2008). *Cell and Molecular Biology.* Wiley.

[NIST and Technology, 1995] NIST, N. I. o. S. and Technology (1995). MNIST database. https://en.wikipedia.org/wiki/MNIST_database. [Online; accessed 16-Jan-2019].

[Read, 2019] Read, J. (2019). Spatial frequencies illusion. https://www.jennyreadresearch.com/misc/illusions/. [Online; accessed 7-Mar-2019].

[Roe *et al.*, 1993] Roe, A., Garraghty, P., Esguerra, M., and Sur, M. (1993). Experimentally-induced visual projections to the auditory thalamus in ferrets - evidence for a W cell pathway. *Journal of Comparative Neurology*, 334(2):263-280.

[Rosenholtz, 2015] Rosenholtz, R. (2015). Texture perception. *The Oxford handbook of perceptual organization* (ed. Wagemans, J.). Oxford University Press, Oxford, UK.

[Tamura and Norgren, 1997] Tamura, R. and Norgren, R. (1997). Repeated sodium depletion affects gustatory neural responses in the nucleus of the solitary tract of rats. *American Journal of Physiology-Regulatory, Integrative and Comparative Physiology*, 273(4):R1381-R1391. PMID: 29586732.

[Taylor, 2015] Taylor, M. (2015). Sdr capacity and comparison (episode 2).

https://youtu.be/ZDgCdWTuIzc.

[Treisman and Gelade, 1980] Treisman, A. and Gelade, G. (1980). A feature-integration theory of attention. *Cognitive Psychology*, 12(1):97-136.

[Wolfe] Wolfe, J. Spatial configuration search dataset. http://search.bwh.harvard.edu/new/data_set_files.html.

Index

索　引

Memorandum

Memorandum

Memorandum

【著者紹介】

Ruggero Micheletto, Ph.D. in Physics
Born in Piedmont in Italy in 1962, he got the degree in physics in 1987 from the University of Torino and the Ph.D. in Physics in 1992 from the University of Bologna, the oldest known university in the world. He moved to Japan in 1994. Currently he is professor of Physical sciences at Yokohama City University. He does research in nano-optics, optical devices and related biological applications, experiment of visual perceptions and neural networks.

Aki Tosaka, Ph.D. in Physics
Obtained Ph.D. in Physics in 2004 from Gakushuin University in Tokyo, Japan. She then was appointed as a researcher in AIST (National Institute of Advanced Industrial Science and Technology) in Tsukuba, IMRAM (Institute of Multidisciplinary Research for Advanced Materials) Institute in Tohoku University, and worked as an assistant professor at Yokohama City University until 2018.

Kotaro Oikawa
Obtained a M.S. (Master of Science) in Semiconductor Physics and Cognitive Informatics in 2012 from Yokohama City University. His current research interest is visual information processing with the aid of computer programing.

* * *

Ruggero Micheletto（ルジェロ ミケレット）
1962年 イタリア ピエモンテ州トリノ生まれ．1987年 トリノ大学で物理学の修士号を取得後，1992年 ヨーロッパ最古の総合大学のボローニャ大学でPh.D.（物理学博士号）を取得．1994年に来日．現在は横浜市立大学学術院国際総合科学群自然科学系列 教授．
研究活動はナノ光学，近接場光学，光デバイスおよび知覚実験，ニューラルネットなど．

戸坂 亜希（とさか あき）
2004年 学習院大学大学院自然科学研究科物理学専攻博士後期課程修了，博士（理学）取得. 産業技術総合研究所および東北大学多元物質化学研究所 ポスドク，横浜市立大学助教を経て，現在は学習院大学やその他高校にて非常勤講師として勤務．専門は表面物理学.

及川 虎太郎（おいかわ こたろう）
2012年 横浜市立大学大学院生命ナノシステム科学専攻修士課程修了．専門は半導体物性科学及び知覚情報科学．現在はコンピュータープログラミングを用いた視覚情報処理分野の研究を積極的に行っている．

英語と日本語で学ぶ知覚情報科学

Cognitive Informatics
in English and Japanese

2020 年 5 月 25 日　初版 1 刷発行

著　者　Ruggero Micheletto　© 2020
戸坂亜希
及川虎太郎

発行者　南條光章

発行所　**共立出版株式会社**
〒112-0006
東京都文京区小日向 4-6-19
電話番号　03-3947-2511（代表）
振替口座　00110-2-57035
www.kyoritsu-pub.co.jp

印　刷　大日本法令印刷
製　本　協栄製本

一般社団法人
自然科学書協会
会員

Printed in Japan

検印廃止
NDC 007.13, 141.2, 141.51, 491.371

ISBN 978-4-320-12460-8